Hubbard, the lawyer, a mineralogist, the writer of this historical memoir,

along with his friend,

Sartor, the "Captain," a Bohemian from San Francisco,

and their

Indian Guides, Silas & Joe,

on ***over a month-long*** journey through the Maine woods.

The 160-mile journey of this tale took place in the fall of ***1881***.

Their mode of transportation were **birch-bark canoes**.

It was when caribou still roamed the Maine woods.

They ate what they carried with them, or what they trapped along the trail; they slept under the stars, and hiked through ten-inches of snow.

It was not an easy trip. Here's what Hubbard had to say:

"The keen enjoyment of many hours had made ample amends for the *few hardships* we had undergone, while the lessons we had had of Nature's teaching will form ***a priceless treasure-book***, of which, when we are far removed from her schoolhouse, we may *turn the leaves anew*, and read again and again the story we had conned."

Almost 140 years later, I give you their story with updated notes and photographs.

A

Burnt Jacket Publishing

Classic Release

# WOODS AND LAKES OF MAINE

A TRIP FROM MOOSEHEAD LAKE TO

NEW BRUNSWICK

IN A BIRCH-BARK CANOE

TO WHICH ARE ADDED

*SOME INDIAN PLACE-NAMES AND THEIR MEANINGS*

*NOW FIRST PUBLISHED*

BY LUCIUS L HUBBARD

COMPILER OF "HUBBARD'S GUIDE TO MOOSEHEAD LAKE AND NORTHERN MAINE"

*NEW AND ORIGINAL ILLUSTRATIONS*
BY WILL L TAYLOR

**2020 ANNOTATED EDITION**
**Edited by TOMMY CARBONE, PH.D.**

**Foreword by Frederick T. Wilcox, Forester**

**THE MOOSE**

**(Alces americanus)**

# WOODS AND LAKES OF MAINE

A TRIP FROM MOOSEHEAD LAKE TO

NEW BRUNSWICK

IN A BIRCH-BARK CANOE

TO WHICH ARE ADDED

*SOME INDIAN PLACE-NAMES AND THEIR MEANINGS*

*NOW FIRST PUBLISHED*

**BY LUCIUS L HUBBARD**

COMPILER OF "HUBBARD'S GUIDE TO MOOSEHEAD LAKE AND NORTHERN MAINE"

*NEW AND ORIGINAL ILLUSTRATIONS*
BY WILL L TAYLOR

BOSTON
JAMES R. OSGOOD AND COMPANY
1884

# 2020 ANNOTATED EDITION

Cover photo, and newly added interior illustrations, and photos from the collection of Tommy Carbone or as otherwise noted.

Cover design by Tommy Carbone.

Cover photo is from the Mt. Kineo fire tower, looking north.

Editor of the 2020 annotated edition, Tommy Carbone.



***Burnt Jacket Publishing***

Greenville, Maine

Library of Congress Control Number: 2020905539

2020-07-19 AMZPBK

ISBN: 978-1-7347358-3-3

Also available in Commemorative Hardcovers:
978-1-7347358-2-6 – Hardcover Dust Jacket Library Edition

www.tommycarbone.com

*

*

**ORIGINAL COVER IMAGE – 1884 EDITION**

# ORIGINAL PUBLICATION 1883

BY LUCIUS L. HUBBARD

§

CAMBRIDGE

ORIGINALLY PRINTED BY JOHN WILSON AND SON

UNIVERSITY PRESS

*

# Introduction to the 2020 Annotated Edition

*

My first visit to Maine was in October of 1993. By December of that year I had relocated and was living in the coastal area of the state. While I enjoy the lighthouses, lobster boats, and rocky shoreline, I've always been more of a woods man, than an ocean and sand goer. It's no wonder many of my Maine vacations and weekend excursions have been to the mountains, lakes, ponds, and bogs.

I first discovered the Moosehead region of Maine in 1995 and I was struck by the ruggedness and beauty of the woods, lakes, ponds, and trails. So it is not surprising that once I visited, I kept returning – for more than twenty years now.

I enjoy clear lakes, where when I stand with water up to my neck, I can see my feet. I am relaxed by the scent of pine blown in on cool mountain breezes. I anticipate being lulled to sleep at night by the wail call of a loon, and woken by the tremolo in the morning.

I have explored the north woods, on foot, by paddle, by oar, and various engines of one kind or another. It is here that I have seen more wildlife than a Brooklyn-born boy could have ever imagined to see up close and actually in the wild. I've stood on mountain tops to see nothing but more mountain tops. It is a place that I treasure being able to call home.

I first became aware of Hubbard's book, "*Woods and Lakes of Maine*," in late 2019. It was recommended to me by a newly made acquaintance, who had found my writings about the outdoors around Moosehead Lake realistic. On his recommendation I tried for months to acquire a copy of the book. All I could find were poor reproduction scans of the original text. Since the work is in the public domain, anyone can copy an old edition of the book and sell it, with little or no regard for quality. Most of these 'publishers' in my opinion were doing an injustice to the value of the text and the subject matter. And while the book is available online to read for free, I wanted a copy to hold in my hands, to have on my bookshelf, to take along on a canoe trip. No matter how I searched, I could find no such volume.

The acquaintance I mentioned has since become a friend, you will find the foreword in this book written by his hand. By Fred finding one of my novels and contacting me, I have learned he too was born in Brooklyn, he moved to the same small town in upstate New York where my family ended up (some thirty years later than he did), and he has spent time in the Maine woods, with a particular appreciation for the forest around Moosehead Lake.

When I mentioned to Fred, I could not find a decent copy of Hubbard's book to purchase, he did not hesitate to send me his copy – it is a treasure of his he wanted to pass down. When I received the book, which was packed more securely than an egg shipped for international passage, I found it to be in excellent shape for an 1884 print edition. However, the pages are yellowing and brittle, and the spine is faltering. When I held the book, I realized I would treasure it, but I in no way would chance bringing it on a canoe trip, or sit on the dock with it to read on a summer evening.

It was then I decided to create an edition of the book I could display, hold, read, and reference without worry of damaging the original. To me, Hubbard's book holds too much historical significance for it to not be available to anyone who wished to have it in their library.

Fred's gift has been invaluable in order to reproduce the illustrations in the original work, as well as to validate copies of the text found online, which have proven to not be one-hundred percent reliable.

It is my intent that the volume you hold in your hands reproduces the original work of Dr. Hubbard, and readers and researchers will appreciate the documentation of an 1881 canoe trip through the Maine woods.

In this edition I have added map pages in the Appendix in order that you may follow the canoe journey of Hubbard, the Captain, and their Indian guides – Silas and Joe. Of course, you can find more detailed maps online if you wish to explore further.

For those interested in the Indian names and pronunciations, I refer you to Appendix I. The diligence Hubbard applied to documenting the Indian names is commendable. His sources are noted in the Appendix.

The book, in the spirit of Thoreau's *The Maine Woods*, includes details of historical significance. Aside from the descriptions of wildlife and the terrain, Hubbard went to great lengths to include original Indian names of places, animals, and objects. He also includes the scientific names of animals. These details provide much more than a memoir of 'roughing it' for a few weeks in the unspoiled Maine forest.

On the facts, distances, and names mentioned in the book, these are all original to the author. I have not, except in situations where noted, corrected or amended his writings.

In this edition, I have made limited corrections to word use and spellings that were common in the late 1800s. Corrections not made are by intent to preserve the historical accuracy of the text. Words, such as, *travellers, marvellous, bevelled,* and *centre* which are now more commonly spelled as *travelers, marvelous, beveled* and *center* are left in the original form.

In limited cases, I have made updates, to more common forms for easier reader experience and typesetting. These include edits that will not detract from the meaning of the text. An example of such a change is the former writing of *today,* as *to-day* in Hubbard's time.

Commas and punctuation have been corrected where necessary and obvious. I can only imagine the great lengths Hubbard had gone through to painstakingly typeset this work in 1884. Even today, with modern word processing, the typing of the Indian names took a great deal of time.

This edited volume includes a brief summary of the admirable life of Dr. Lucius Lee Hubbard. Discovering facts about the author, while working on this edition, was fascinating to me.

He was a husband, father, lawyer, and then in pursuit of his fascination with minerals he earned a Ph.D. in minerology from the University of Bonn. He had a long and distinguished career in geology and academia. His studies and research had him traveling extensively. Amazingly, with all his endeavors he found time to write not one, but two books on the Moosehead

Lake region. He led a very fulfilled life. His inquisitive nature and his desire to share with others his learnings is evident in his writings.

While this book describes a 160-mile journey, over 130 years ago, I am certain that there are places described in this text that have never felt a man's boot. These places are as natural as in the day Hubbard passed through the region. And while much of the north woods remains the same, there have been changes in the last century to the Maine landscape. It is with respect to those differences from the time of the author's writing, that annotations in this edition have been added. Additionally, new photographs are included where they fit with the topics of the chapters. All new additions are marked with a "**2020**" to clarify original versus new notes and photos.

As you read, you may wonder how many highways or grocery stores an explorer would find should they attempt to follow Hubbard's original journey. I am happy to report that Maine's north woods are much the same as they were in 1881. Even now, in 2020, most of this area is without cell service, so if you go, be prepared for remoteness.

Greenville, Maine, where now the roads are paved, still has not one traffic light (as of this writing). The steamer, Katahdin, while different from the one Hubbard traveled on, still makes trips up the lake in the summer months, but with no destination to unload passengers other than where they started. The Kineo hotels are no longer, and the sporting camps along the lake are now reached by pickup, BMW or SUV, not train or steamer.

Thankfully, there are no highways that venture along the trail of this journey. As for grocery stores, you will need to pack what you need and carry out what you don't use, for along the Hubbard trail there are few supply points.

It is my hope that this new edition will provide you, the lover of Maine, the North Woods, or the outdoors in general, with reading pleasure as it has done for Fred and I.

I have found this text to be a timeless piece of wilderness writing. I am pleased to make available this updated edition with notes and references about the North Woods of Maine.

Tommy Carbone, Ph.D.

2020 Edition Editor of, "*Woods and Lakes of Maine*."
Greenville, Maine
February 28, 2020
www.tommycarbone.com

*On my social media sites and blog, I share photos and information about the Moosehead Region. Please follow and stay in touch with me there.*

# Foreword

*

March 1, 2020

*

In the fall of 1957, I arrived at the University of Maine in Orono as a young forestry student. In our studies it wasn't long before we attended classes and field trips at the University Forest and the Caribou Bog near Old Town. Being from the lower Catskill region in New York, which was mainly a deciduous forest region, I had never observed or set foot in a northern coniferous forest or a sphagnum bog and at the time I thought this was the most beautiful environment I had ever set foot upon. Having always had a deep interest in local flora and fauna I took advantage of every opportunity to explore areas beyond the vicinity of the University. Without a means of transportation, it wasn't until my second year of study that I had the opportunity to visit the area far to the North and West of Orono. Then I made my first trip to the Moosehead Lake region. Until then I never realized there were so many large wilderness lakes in existence and I quickly fell in love with this area.

It wasn't until after I graduated from the University and married that I began my exploring, camping, and canoeing in the Moosehead Lake region. Over the years I paddled hundreds of miles on places such as the West Branch of the Penobscot, Moosehead Lake, the Roaches, Ragged Lake, Brassua Lake, Chesuncook Lake, and Lobster Lake. On each and every trip I was fascinated with the trees, flowers, wildlife and everything a forester would love. It was here in Maine that I discovered my favorite habitat, a slow-moving stream surrounded by bog and covered with American Larch, Black Spruce and Labrador Tea, and home for beaver and moose.

It was in my journeys that I came across places and names such as Kineo, Katahdin, Mooselookmeguntic, and Seeboomock. Having a deep interest in our Eastern Native Americans and their culture, I was fascinated by the names given to the mountains and lakes and I immediately began a search of a publication that would help me with the meaning of these names. Through my search I discovered there was a book titled, *"Woods and Lakes of Maine,"* written by Lucius Hubbard, which contained a section, SOME INDIAN PLACE-NAMES AND THEIR MEANINGS. However, since the book was published in 1884, I realized that finding a copy could be a difficult task. Fortunately, I found a copy rather close at hand and soon this book was in my possession where it remained for almost 50 years.

When I began to read this book, I discovered it was far more than a list of Indian names and their meaning. It describes in infinite detail, Hubbard's sojourn from Moosehead, across the "North Woods" to Canada. It's a detailed account of traveling by canoe and surviving in the wilderness. There are beautiful descriptions of the forests, the lakes, the streams and the wildlife.

Nearly 150 years have passed since this book was written and while much has changed, this section of Maine has largely remained the same. If one was to make this journey today,

Hubbard's descriptions of the trip would largely remain the same as detailed within this book.

Unfortunately, Caribou are no longer a part of this locale. One will also note within this book some names for trees, birds and animals which are not commonly heard these days.

I have canoed over a portion of this trip with modern gear such as a light canoe, pack, and tent as well as the modern convenience of freeze-dried foods. It was actually this book that inspired me to paddle along the Penobscot and Lobster Stream to enter Lobster Lake. While there I made a decision to continue to Little Lobster. However, there were so many beaver dams along this route that the trip proved to be too much of a challenge. What hasn't changed over the decades is the time and effort it takes to portage from one body of water to another.

Being a professional forester and an avid lover of the outdoors, Lucius Hubbard's book, *"Woods and Lakes of Maine,"* greatly assisted me in my professional career and my love of the outdoors. Dr Hubbard not only observed trees, birds, fish and mammals but he studied and described them in great detail. This was not only in appearance, but more importantly, their movements, their habits, their sounds. This is something I have learned to do over the years. During the later months of each year, I count and report on migrating raptors. I have learned it is much easier to identify birds, not just from their physical appearance as described in a field guide but from the way they hold and flap their wings, their movement in flight, the position of their legs. It's little details which helped me make the correct decisions in my career as a forester.

I am very happy to learn that Dr. Carbone has made available an updated version of this great and beloved book. My original copy of the book is again in good hands. This new version will now take its place on the book shelf and will be available to those who love the North Woods as much as I do.

Frederick T. Wilcox
Forester, retired

# Dr. Lucius L. Hubbard

## 2020 Edition

Lucius Lee Hubbard
1849–1933

Lucius Lee Hubbard was born at Cincinnati, Ohio, August 7, 1849. He was the only child of Lucius Virgilius and Annie Elizabeth (Lee) Hubbard. He was born following the passing of his father, who was an 1824 graduate of Harvard.

He attended Phillip Exeter Academy for two years, and graduated from Harvard in 1872. Following college, he traveled abroad studying the German language, history, and international law at the University of Bonn.

In 1875 he received his Legum Baccalaureus. LL.B., *Bachelor of Laws* degree from Boston University. This degree was almost universally the first professional law degree in the United States. He subsequently was admitted to the bar.

On September 29, 1875, he married Frances J. Lambard of Augusta, Maine. Until 1883, he resided in Cambridge, Massachusetts. Outside practicing the law, his interests were varied. He was an avid collector of stamps and he owned one of the largest collections of early Americana, including original Robinson Crusoe books. Ultimately, his personal fireproof library became too small to house his stamp and book collection and many were sold.

He had an affection for minerals and the outdoors. In 1877, he published a small book, entitled, "*Summer Vacations at Moosehead Lake and Vicinity.*" This became a starting point for a number of books about northern Maine, which he compiled and published as, "*Hubbard's Guide to Moosehead Lake and Northern Maine.*"

In the fall of 1881, Hubbard, along with a personal friend and two Indian guides, embarked on a journey of over 160 miles in a birch-bark canoe; that trip is the subject of this book, "*Woods and Lakes of Maine.*" Over the years, he would return to the Maine wilderness that captivated his heart.

He published, "*Map of Northern Maine – Specially adapted to the Uses of Lumberman and Sportsmen*." His map was of significant detail for the period. He updated and re-issued it several times. An image of the 1899 edition of the map is included in this book. (You may find versions of the map online where you will have the capability to magnify areas of interest.)

Hubbard was particularly interested in the rocks of Mount Katahdin. His interest in minerals once again brought him to Europe, where in 1883 he studied at the Bonn, and earned a Ph.D., with a mineralogical focus. Following his degree, he traveled for research to Heidelberg (Germany), Switzerland, and Italy. By the summer of 1887 he was spending time again in Maine.

In 1890, Dr. Hubbard was drawn away from New England, when he accepted an offer to join the staff at the prestigious *Michigan Geological Survey and School of Mines*. He later became a Regent of the University of Michigan. Interestingly, it is reported that Hubbard was partially colorblind, a disadvantage that did not alter his ability to be a successful mineralogist. Over the years he had collected a substantial mineral collection, which he donated to the College of Mines.

From 1893 until 1899 he was a Michigan State Geologist. A position he resigned from due to unsatisfactory provisions from the State for Geological Survey work. This was probably most beneficial, for following his resignation, he worked for various mining concerns and within that same year, he had located the Champion Mine. Minerals found there included copper crystals, prehnite, and chalcocite. This mine became one of the most productive in the States, and material was mined there, on and off, through 1967.

Dr. Hubbard did not sit idly in retirement. From 1905 through 1917 he was an appointed member of Board of Control of the Michigan College of Mines. He was reelected continuously to the University of Michigan Board of Regents from 1910 until he resigned in the year of his death (1933) due to health reasons.

He celebrated over fifty years of marriage to Frances J. Lambard, until her death, several years before his passing. Together they had four children. Later in life he spent his time between winters in Florida and summers on the Keweenaw Peninsula in Michigan.

Dr. Lucius Lee Hubbard passed away in 1933, at Eagle Harbor, not far from the Champion Mine. He is buried in Forest Grove Cemetery, in Augusta, Maine.

*This composite of Dr. Hubbard was obtained from the following sources by the editor.*

1. A presentation at the fourteenth annual meeting of, ***The Mineralogical Society of America***, Chicago, Illinois, December 28, 1933.
2. ***Seventh Report of the Secretary by Harvard Colleg***e, on the Class of 1872.

# Preface to the 1888 Second Edition

**In** offering to the public the second edition of "Woods and Lakes of Maine," the author makes his sincere acknowledgments for the generous meed of praise bestowed upon the first edition by the press, for the many kind words of approval given by personal friends, and for the unexpected, and hence all the more gratifying testimonials received from utter strangers. These are an ample reward for his labor, and are the best evidence that he had attained his object, — that of keeping fresh in the minds of his readers pleasant memories of their outdoor life. He is thus led to hope that this new and revised edition will find favor equally with the other, at least among those whose privilege it is, at some season of the year, to commune with Nature in the wild recesses of the forest.

Boston, May, 1888.

# Introduction (Original Edition)

**It** had been the writer's wish for several years to give to the public a true and circumstantial delineation of the camper's life in the Maine forests, especially as seen by one who goes into them with an Indian guide. The gratification of this wish was not possible until the autumn of 1881, at which time it was the writer's good fortune to be accompanied on his annual excursion into the woods by the friend to whose brush the illustrations in the following pages are due. The tendency to sacrifice literalness and accuracy to artistic effect prevails to a great extent today among professional illustrators. Moreover, they do their work principally in the studio, from photographic material, it may be, and often have to draw on their imaginations to fill up gaps here and there, or to supply some deficiency left by the camera. Their pictures of forest life, therefore, are apt to convey, if not an incorrect impression, at least an in harmonious or incomplete one, lacking color and reality, detail and finish, qualities the ability to produce which can only be acquired by seeking Nature in her wildest haunts, and drinking at the fountain-head.

The excursion above referred to has been made the groundwork of the accompanying text, and a few experiences and observations borrowed from other excursions of the writer have been introduced to make the work more comprehensive, and, it is hoped, more interesting as well. While no attempt at exhaustiveness has been made, the writer's aim has been to introduce largely such features of forest life as he had not happened to see treated before in print.

It must not for a moment be supposed that the doings and sayings of the writer's friend and companion are herein literally portrayed. A foreground was wanted, to complete the picture, which should be in contrast with the subdued tones produced by the usually dull routine of camp life; and accordingly a mythical character was evoked, who should consent to play the clown, and to be laughed at for his wit or his stupidity, as occasion might require. (1)

Another reason for this publication besides the one offered above is the writer's wish to make known a number of Indian place-names, and several legendary traditions, which he has gathered from Indian sources during his vacation rambles through Maine. The growing interest manifested of late in this subject leads him to think that the publication of these names and their significations, although in the latter no pretension is made to philological precision and fulness, may be received with favor, and may lead to more thorough researches in the same direction by others. The translations (in many cases merely explanations) of the Indian names, as given in the foot-notes, have been derived, principally if not wholly, from the Indians themselves, and generally in the very words quoted. In the Appendix an attempt has been made to enlarge somewhat the sphere to which the writer limited himself in the text, by comparison, as well as by analysis, the latter often fragmentary to be sure, and sometimes offered with diffidence.

The accompanying map*, although a few copies of it have been heretofore published, was revised especially for this book, and contains with but two exceptions all the Indian names of places herein referred to that lie within the district covered by it.

That our wild forests, and the lakes and streams which fill their basins and crevices, as it were, contain a vast fund for man to draw on, a fund of all that is precious to health and recreation, and an inexhaustible mine for study and investigation, needs no argument. In subordinating to these opportunities for pleasure the more exciting sports of forest life, the writer would not be understood as decrying or detracting from the latter. He would merely give deserved pre-eminence to that more lasting pleasure, of drawing from Nature the bounties which she offers in profusion, of learning to read her stones, her leaves and blossoms, and of forming a nearer kinship with the wild offspring that swim in her waters or roam through her groves. He has touched lightly upon these boundless themes, and if his words shall bring pleasure to any who have trodden the path before him, or incite others to follow, his reward will be ample.

Cambridge, Mass., 1883

***2020 Edition Note** – the map referred to was an insert provided in the original printing. An image of the updated map is included in the appendix. For an image of the map that can be enlarged, please search online.*

(1) The companion Hubbard makes reference to in this book as, "The Captain," was the famed illustrator William Ladd Taylor (1854-1926).

Original Dedication

*To her*

FROM WHOM HE DREW THE LOVE OF NATURE THAT IN HIM LIVES.

*THE WRITER*

AFFECTIONATELY DEDICATES

*This Book.*

## About

William Ladd Taylor

(1854-1926)

Taylor was born in 1854 in Grafton, Massachusetts. He attended Worcester public schools and worked for a time as a draftsman. He studied art in Boston, New York City, and later Paris. Taylor's paintings and illustrations depict a wide array of American scenes in which he sought to preserve the past.

He had a summer home on lake Mooselookmenguntic in Maine.

*

# Contents

# Detailed Chapter Topics

**CHAPTER I.**

Moosehead Lake. — Its Location. — Altitude.— Extent.— Depth.—Indian Name.—Legends.—Modern Name.—Mount Kineo. — Spencer Mountains.— Indian Fights.— Indian Place-Names.

**CHAPTER II.**

Passage over Moosehead. — Our Party. — Joe and Silas. — Northeast Carry. — Old Tramway. — First Whiff of Forest Air. — West Branch of the Penobscot. — Lobster Lake. — Moonlight Reveries.

**CHAPTER III.**

Penobscot Valley. — Rapids. — First Morning in Camp. — Breaking Camp.— "Pitching" and Loading Canoe. — A Caribou. — Measuring Distances. — Fox Hole. — A Mess of Trout. — A Mishap. — Reflections. — Running Rapids. — A Fifth Passenger. — Squirrels.

**CHAPTER IV.**

Preparations for Night. — The Captain's Opinion on Camping. — A Misty Morning. — Chesuncook. — Up the Umbazookskus. — Smith's "Jumper." — Its Effect on Moosehead Guides. — Making a Portage. — Mud Pond Carry. —Native Modesty.

**CHAPTER V.**

Black-Ducks. — A Muskrat House. — The Muskrat as a Pet. — Chamberlain Farm. — Apmoojenegamook. — Into Pongokwahemook. — Evidences of Illegal Hunting. — Some Reflections on Game Protection and the Game Laws.

**CHAPTER VI.**

The Loon considered Musically and otherwise. — Camp on Nahmajimskitegwek. — Hornets and Maple-Sugar. — Visitors.— An Excursion to Haymock Lake.— Rough Water.— Allagaskwigam'ook. — Through the Breakers. — Guides. — Indians vs. Whites.

# List of Appendices.

# List of Original Engraved Illustrations

** Engraved by John Andrew & Son.*

# 2020 Edition Photographs

# Chapter I

*

Moosehead Lake. — Its Location. — Altitude. — Extent. — Depth. — Indian Name. — Legends. — Modern Name. — Mount Kineo. — Spencer Mountains. — Indian Fights. — Indian Place-Names.

*

**MOOSEHEAD LAKE**, the largest of two or three thousand lakes and ponds[1] with which the State of Maine is dotted, lies just above the west central portion of the State, approximately between parallels 45° 25' and 45° 50', and meridians 69° 30' and 69° 48', and is a vast reservoir, whose waters are used extensively to float to market the yearly timber product of the Kennebec valley, and to furnish motive power to the many mills along the river's course. The lake, according to barometrical computations[2] made by the writer in 1881, is 995 feet above mean tide, while according to other estimates made during the present century its altitude is variously stated to be between 960 and 1060 feet.[3]

From Greenville, a small village at its foot, and the only one on its shores, the lake extends some thirty-six miles to the north;

---

[1] **2020:** There are officially over 2,400 named lakes and ponds with a surfaces area of 10 acres or more wholly in Maine. There are over 2,600 additional lakes and ponds with no name and another thousand, give or take, that are under 10 acres.

[2] From eight observations, made on three successive days in August, simultaneously with observations at Orono by President M. F. Fernald of the Maine State College.

[3] **2020:** Current valuations are 1,027 feet (313 meters) above sea level.

and varies in width from 194 rods,[4] opposite Kineo Promontory where the telegraph-cable crosses, to about twelve miles, the distance from the east or principal outlet to the head of Spencer Bay. Its outline is remarkably irregular, and readily suggests the ravines and other depressions between bits of high ground, up into which a stream, checked by a barrier of Nature's sudden interposition, might have flowed back, turning former ponds into bays, and making islands of hills and promontories of ridges. The area of the lake is said to be 120 square miles,[5] and its shore line at high water about 400 miles; but these latter figures are evidently much too large.[6,7]

Fabulous stories have been told of its depth, principally on the authority of men who have towed logs over its surface, and their measurements vary from five hundred to a thousand feet, the latter depth being claimed between Farm Island and the Shaw Place east of it. Seventy-five soundings made by the writer in 1881 and 1882 failed to disclose a greater depth than 231 feet.[8] The character of these soundings was such as to make it probable that in the middle parts of the lake the bottom is uniformly level. The inclination of the bottom from either shore

---

4 **2020:** A rod is English measure of distance equal to 16.5 feet (5.029 meters).

5 Wells's Water Power of Maine, (Augusta, 1869,) p. 98.

6 See Appendix III. 8 See Appendix IV.

7 **2020:** Today's accepted metrics for Moosehead Lake put it at approximately 40 miles long by 10 miles wide, with an area of 120 square miles, and a shore length of 280 miles. Hubbard was correct that the early estimates put the shore length at too high a figure.

8 **2020:** Current max depth estimate is 246 feet, remarkably close to the author's 1881 measurement.

to the middle parts is not alike on all sides, the descent being, in some places, very gradual, while in others it is quite rapid.

The only theory on which the existence of these reported depths can rest is, that there are isolated holes or narrow rifts in the underlying rock; a theory that seems quite improbable, as the ceaseless movement of the waters and the accumulations from incoming streams during ages past must long ago have filled any such places, if they ever did exist.

The name by which Moosehead Lake is known to the Penobscot Indians is *Xsébem'*, while to that remnant of the Abnakis that have lived at St. Francis it is *Sébam'ook*, either of which is freely translated by "extending water," the second form having the locative ending.[9] On John Mitchell's Map of North America, published in 1755, Moosehead Lake is called *Chenbesec L.* The application of this name was probably made by mistake to Moosehead Lake instead of to Chesuncook Lake, and two facts strengthen this probability. First, the Indians of today have no form of word like it for the first-named lake, but unite in giving other names which *inter sese* are practically identical, in meaning at least. Secondly, on Governor Pownall's map, published in 1776, *Chesuncook* is called *Chenosbec*. The three forms *Chenbesec, Chenosbec*, and *Chesuncook* are probably variations of one word. According to the Penobscot Indians,[10] *Chesuncook* means "the biggest lake" (i.e. on the

[9] A St. Francis Indian once told the writer that Xsebem' would naturally and properly be the exclamation of one of his tribe when, going through the forest, he should suddenly see light ahead through the trees, and the sheen from an open body of water. Cf. ωasséghen (Râle's Abnaki Dictionary).

[10] John Pennowit and others. See Appendix, Indian Place-Names

Penobscot); but the word has thus far apparently defied all attempts by Indian scholars at analysis, unless its middle component, *sunc*, be the equivalent of sa$^{n}$k, "an outlet," in which case the meaning is "great outlet place."

Again, on Pownall's map, a lake much smaller than *Chenosbec*, but with a large island in the centre, and evidently meant for Moosehead Lake, is called *L. Sebaim*. On Andrew Dury's Atlas, published in London in 1761, on Sayer and Bennet's map,[11] published in 1776, and on that of Thomas Jefferys, published in 1778, the same lake is called *Kese-ben. Sebaim, Keseben,* and the more modern *Xsébem'*, or *Sébam'(ook)*, may safely be said to be identical. The form nearest them appearing as a substantival component in Maine place-names is *pegωâsebem* (Râle) or *quasabam*, "lake." The absence of any of the forms of *Xsébem'* as distinctive name for any other body of water in that State furnishes very good evidence that it was not simply a generic term, but was in and of itself a name given by the Indians specifically to Moosehead, and so understood by former map-makers.[12]

---

[11] Map of the Province of Quebec, according to the Royal Proclamation of 1763.

[12] Just when the present English name of the lake was first used does not appear. On Jay's map, published in 1786, we find it called *Moose Lake*, and on plans of more recent date we find *Seboumook* (Greenleaf's map), *Se-baumock, Seebommock, Seeboumock,* and *Seboumock*. (See Plan Books, B.1, PI.27, B.2, PI.9, B.7, PI.42 and 57, 1814-15, Land Office, Augusta.) Sullivan calls it "Moose Pond or Moose Lake " (History of the District of Maine, Boston, 1795, p.33). Montresor called the lake "Orignal,"which is the French for "*moose*." The statement in the text should be taken in connection with the explanation under "Sebec" in the Appendix.

The origin of the name Moosehead Lake is probably due to one of several old Indian legends, which associate the place with adventures in which the moose bore a prominent part. One of these traditions says that "in the olden time men and animals grew to an immense size. The Indians thought the moose were too large, and sent a hunter to make them smaller. He killed a big bull, Kineo Mountain, and reduced his size by cutting slices from his body. The rock at the foot of the mountain today looks like steak; streaks of lean and fat can be plainly seen in it. The hunter cooked his meat, and afterwards turned his kettle, Little Kineo Mountain, on its side, and left it to dry. So, the moose grew smaller and smaller."

**POLING UP THE RAPIDS**

*

This legend, in almost the exact words given above, came indirectly to the writer from Louis Annance, then an aged Indian, who had been educated at Hanover, New Hampshire, and was afterwards "Sangamon," or chief, of the St. Francis tribe, and later an inhabitant for ten years of the Moosehead forests. He is no longer living.

Another legend says that there was an old Indian, who, in the words of the writer's informant, "was chief of the whole nation. He was capable and could do anythin', same as God, — make anythin'." While on his way through the forests, one day, he came upon two moose, hurriedly dropped his pack, and started in pursuit of them. The smaller moose, Kineo Mountain, was soon overtaken and killed. The chief, after boiling some of the meat, turned his kettle upside down, so that it should not rust, took up the trail of the larger moose, and followed the latter down to Castine, where he killed and dressed it. The heart, liver, and other entrails, he threw to his dog, and they are the long string of rocks which are there to this day. The more easterly of the Spencer Mountains is *Sabōta'wan*, "the pack," while the other, or western, peak is *Kōkad'jo*, "the kettle."[13]

The authority for this latter legend is John Pennowit, an Indian of the Penobscot tribe, who has passed the greater part of eighty-eight years in the woods of Maine, to whom the writer is indebted for much of the information on Indian nomenclature contained in these pages. He is probably the same man

---

[13] The writer has not been able to find any modern Indian name for "pack" like *Sabōta'wan*, but *Kōkad'jo* is formed from *kōk*, "kettle," and wadjo, "mountain." *Sabōta'wan* is said to mean more exactly "the end of the pack, where the strap is pulled together."

incidentally mentioned by Thoreau as John Pennyweight. The writer has no hesitation in accepting, as the more correct, his version of the story which establishes the identity of West Spencer Mountain with the "kettle." Indeed, the shape of the Spencers would seem to settle the matter beyond question. *Kōkad'jo* is quite round, while *Sabōta'wan* is long, and its top level, the eastern end being squared off much like the end of an Indian pack.[14]

These traditions show that the Indians were endowed with great imaginative powers and with no little poetic feeling. Mount Kineo, when seen from the southern side, looks not at all unlike an immense moose, lying or stooping with its head towards the west. The precipitous eastern cliff is a very good counterpart of the rump, while a slight elevation at the beginning of the western slope well represents the withers, and another near its foot the swelling of the nose or "mouffle." Indian imagination, however, did not stop here. The two main arms of the lake, which extend north and south, one on each side of the "moose," with their numberless bays and coves, form the animal's antlers with broad blades and branching prongs. May not this be the origin of "Moosehead?"[15]

---

[14] Another Penobscot Indian adds to the foregoing story the statement that this great chief, when about to overtake his moose at Castine, jumped from Belfast over to Castine, and left the trail of his snow-shoes on the side of a hill near Belfast, where it can be seen to this day. This is called Mada$^{n}$gamus. A high ledge on the opposite side of the bay is called Moosay-eek-teek, the "moose's hind-quarters."

[15] Since the above was written, the writer has found the following on page 10 of Hugh Finlay's Journal, published in 1867, at Brooklyn, by Frank II. Norton. Finlay was a surveyor, and in September, 1773, with four Indian guides from the Chaudière, he journeyed through the

The position of Mount Kineo, on a slender promontory midway of the lake and at its narrowest part, is very marked. It is one of a broken chain of small mountains which extend from Lobster Lake on the northeast across Moosehead to Blue Ridge. According to Dr. Jackson its formation is hornstone,[16] and it is said to be one of the largest masses of that substance in the world.[17]

*

**2020 Edition**

**MT. KINEO**

*

The word *kineo* is said to be Abnaki for "high bluff," and is a very good description of the mountain as seen from the south or east, on which sides it is a sheer wall of almost bare rock,

---

Maine woods and over part of Moosehead Lake, having entered the latter at the northwest arm, from Carry Brook. He says: "We march'd thro' the woods, a mile S. to another dead creek half a mile in length leading us also S. winding to a large lake called by our Indians Moose-parun. . . . . This lake takes its name from a very remarkable mountain on the S. side, about nine miles down; the Indians say it resembles a moose-deer stooping."

[16] Geology of the State of Maine, by C. T. Jackson

[17] **2020:** The rocks that form the cliffs of Kineo, are now generally known as “Kineo rhyolite,” a volcanic rock rich in silica. The high silica content results in conchoidal fracturing, which makes the rock easier to shape. Source: Maine Geological Survey, 2018. Image included.

rising to a height of seven hundred and sixty-three feet above the lake. This mountain was in olden times a place of great resort for the Indians, who went to it from their distant villages to get its flinty rock to make into arrowheads and other implements. Numbers of these relics of a past generation have been found, both near the mountain and far away from it. The identity of the rock from which they had been made seems to have been fully established.

*

**2020 Edition:**[17]
**SHAPED RHYOLITE**.
Found near Sandbar Island, Moosehead Lake, Maine.

*Picture courtesy of Frederick T. Wilcox.*

*

From the top of Mount Kineo a fine prospect opens before the beholder. At his feet the lake stretches its far-reaching arms in almost every direction, and in its very silence is awe-inspiring.[18] Beneath him juts out into the water a broad tongue

---

[18] **2020:** The cover photo was taken by the editor from the fire tower near the summit of Mt. Kineo. The view is looking north towards Northeast Carry. The first Kineo tower, made of wood was installed in 1910. It was replaced with a steel tower in 1917.

of level land, on the end of which, in a white cluster, stands the Mount Kineo House[19] with its dependencies. The surface of this ground is gravel, which, in accordance with a theory of ten years ago, might have been deposited by an ocean current from the north. The deposit may, however, be due to glacial action, and contains several depressions, like glacial sink holes. One of these is now a secluded little cranberry-bog.

*

**2020 Edition**

**MT. KINEO HOUSE**

*

From contemplation of the waters, edged for miles with the unbroken forest green, the eye seeks the mountain tops. Kōkad'jo,[20] on the east, rises in a compact rounded mass to the height of 3,035 feet;[21] while Sabōtá'wan,[22] its neighbor, from this point cone-shaped to the view, is perhaps a hundred feet

---

[19] **2020:** In 1881, Hubbard would have been referring to the second Mt. Kineo House. It was expanded in the spring of 1882 and was destroyed by fire later that fall. Variations of the hotel operated through the early 1900s.

[20] **2020:** Kōkad'jo – Little Spencer is 2,992 ft. Kettle Mountain.

[21] From observations made with Green's mountain barometers, by Dr. J. J. Kirkbride and the writer, in September, 1882, the mountain being 2,022 feet above Spencer Pond, and the latter 18 feet above Moosehead Lake.

[22] **2020:** Sabōtá'wan – Big Spencer is 3,230 ft.

higher.[23] Ktaadn[24] [Katahdin] lies beyond them, almost entirely shut out from sight. Farther to the south are the Lily Bay Mountains and turreted Baker in one almost indistinguishable mass, the latter's highest peak being 3,589 feet above sea-level.[25] East of them White Cap[26] and other peaks near Katahdin Iron Works[27] are visible. Squaw Mountain[28,29] is the most conspicuous height south of the lake, its altitude being 3,262 feet. The extent of one's vision on the east and south is limited; but as the eye sweeps the western horizon on a clear day, mountain after mountain comes into view, like the rolling billows of the ocean. The most prominent of these masses are cone-shaped Bigelow on the southwest, and Bald Mountain[30] on the northwest.

Tradition makes Moosehead Lake the scene of many fierce encounters among the Indians, in which the invading Maquas, or Mohawks, took a prominent part. Indeed, the ancient name

---

[23] Estimated from the summit of Kōkad'jo by the aid of a pocket-level.

[24] "The biggest mountain."

[25] From observations made with Green's mountain barometers, in August, 1882, by the writer and an observer at Katahdin Iron Works.

[26] Wassum'ke'de'wad'jo, White Sand Mountain.

[27] **2020:** Katahdin Iron Works (KIW) was likely of interest to Hubbard, as it was there where iron ore was mined from 1843 – 1890. A single iron kiln remains on the State historic site.

[28] A legend concerning Squaw and Kineo Mountains is omitted here because it has been previously printed in the writer's "Guide to Moosehead Lake and Northern Maine," and because there are some doubts about its being Indian.

[29] **2020:** The use of the word, "squaw" was removed from Maine places in the year 2000. It remains here as the original text. Debates on the Indian term for woman and/or body parts are left to other scholars. See also Appendix I for definition. Squaw Mountain is now Big Moose Mountain. (Source: Legislative Document, H.P. 1712., January 10, 2000.)

[30] Eskwe'skwéwad'jo, She-Bear Mountain

of Wilson Pond is said to be Étas-i-i'-ti, "where they had a great fight," or "destruction-ground," and many were the arts resorted to by contending foes to gain the advantage over one another. The Mohawks were persuaded by the colonists to join them against the Eastern Indians, and appear to have been a great scourge to the Penobscots, who were finally compelled to buy peace at the price of an annual tribute, which the Mohawks for many years collected with great regularity.

* * *

Among the ponds and streams near Moosehead Lake, or which empty into it, and whose Indian names are still known, may be included Roach Pond, or Kōkad'je-weem-gwa'sébem, Kettle Mountain Lake; Spencer Pond, the diminutive Kokad'jé-weem-gwa′sébem, Kettle Mountain Pond; Indian Pond on the Kennebec, or See-bah'-ti-cook, Logon Stream; Moose River, or Sahk-ha'- bé-ha-luck', so called because "there is more water flowing from it than from any other stream that empties into the lake"; and Brassua or Brassaway Lake, Psis-con-tic,"handiest place to build canoes"(?). An explanation of the word "Brassaway" is that it is the corresponding Indian word for "Frank," and that the lake has taken its name from some noted chief who lived or died there. It may not be out of place here to say that Indian place-names almost always describe some physical characteristic, natural or accidental, a concomitant of the place or its neighborhood, or a fancied resemblance to something. Sometimes they describe an event, or series of events, that have happened at the place. Less frequently they are applied to objects as subordinate to or near some more prominent object, whose name they embrace as a component. Again, we occasionally find names of persons transferred to places. In the latter class the substantival for lake, river, etc. is generally wanting, and in such cases, we may depend upon it that the names are not of Indian application. Examples of this class are Brassua, given above, and Atean (Pond), the names of chiefs. A like instance, not of a proper name however, is that of Tomhegan (Stream and Pond), from tomahhégan′, which means "axe" or "hatchet." No Indian would think of calling a stream "hatchet." He might say "at the hatchet-stream," or "at the hatchet-rock," and refer to the loss or discovery of a hatchet at that place, or to a rock shaped like a hatchet. At the end of place-names too, we frequently find a locative particle, in k; which serves to change an indefinite or general name into a particular one, or to locate it specifically. The Indian says "at Gull Lake," where we say simply "Gull Lake." The reason for this is more apparent when we remember that a stream, for instance, need not, and seldom does, bear throughout its course the characteristics that its name indicates. Indeed, it is not unusual for a river to have several names, which vary with its characteristics. The name then, instead of being the name of the river, is really and primarily the name of a place on the river, or near it. A lake, too, is sometimes named from its outlet.

One of the prettiest streams that empty into Moosehead Lake is the Soca'tean. Its name is currently thought to be that of a former warrior, Soc Atean, "Standing Atean," who may have had this sobriquet given him for bravery. It is quite probable, however, that the word is an abbreviation or corruption of the real name of the stream given by Pennowit, Mésak'kétéságewick, and explained by him as "half burnt land and half standing timber, with the stream separating them." The syllables sak-ké-té bear a striking resemblance to the modern name. While the name "Atean" exists to this day among the Penobscot Indians, careful inquiry by the writer has failed to find traces of it in combination with "Soc." Perhaps then soc or sakké (Râle) refers to the "standing" timber.

# Chapter II

*

Passage over Moosehead. — Our Party. — Joe and Silas.— Northeast Carry. — Old Tramway. — First Whiff of Forest Air. — West Branch of the Penobscot. — Lobster Lake. — Moonlight Reveries.

*

**ON** a bright September day in the year 1881 the writer found himself on board that asthmatic little craft known to all frequenters of the Mount Kineo House as the "Day Dream,"[1] steaming rapidly around Kineo Promontory towards the northeast corner of Moosehead Lake. His companions were a friend, Captain Sartor, and two Indian guides, Joe and Silas. Two canoes, one of birch- bark and the other of rigid canvas, were firmly lashed, bottom outward, one to each side of the little steamer, while on the latter's deck were piled, in disorder, camp equipage, and provisions estimated to last four men for twenty-five days.

Captain Sartor, although he hailed from San Francisco, was a cosmopolitan, or, as he himself suggested, a Bohemian, and a man of much experience, except in those matters that pertain to life in the woods. He could converse well on many subjects, was at home in the use of pencil and brush, and, although frequently given to scoffing, was endowed with an evenness of temper and a willingness to bear discomfort that made of him that rarest of persons, an agreeable travelling companion. In

---

[1] **2020**: The Katahdin is the only remaining steam vessel operating on the lake.

figure he was large and stout, in language sometimes forcible, but in habits his more genial friends thought he was abstemious to a fault.

*

**2020 Edition**

**KATAHDIN STEAMSHIP**

*

Joe, the elder of the guides, had formerly lived on the banks of the St. John in New Brunswick, and later in Quebec. He belonged to the Maliseets, an off shoot of the Passamaquoddies. Upwards of fifty years old, he was rather lightly built, but was tough and sinewy, and utterly devoid of any tendency to laziness, a trait too often found among his race. A man of great self-reliance, he was in the habit of taking the lead at all times, and by so doing in our party on more than one occasion he evidently wounded the susceptibilities of his younger companion, who was also a man of energy and of great experience in the woods. Joe was very intelligent, and what learning he possessed had been acquired under difficulties. To quote his own words, he said: "Since I come over from Canada I've been study the lon-gwage, and now I kin buy my own grub, and write my own letters. I done pretty well for me. I never was in English school. I have eight children when I first come into Maine. Nine weeks afterwards my wife died. I call it devilish hard time — that *time*. I work the day and study the evenings."

Joe was very well informed about the Northern Maine woods, but once or twice when we called for his estimate of distances and other measurements, and then questioned their accuracy, he seemed to become irritated, and asked us why we sought his opinion if we already had the information demanded.

Silas, an Abnaki of the St. Francis tribe, was thirty years old, and had spent the greater part of his early life as a hunter and trapper in the Canadian forests. His father, P. P. Wzokhilain, had been educated at Hanover, New Hampshire, and had published several books in the Indian language.

The son in order to escape a compulsory education, enlisted during the late civil war, at the age of fourteen, as a private in a Michigan regiment. Three times rejected on account of his youth, he was finally accepted on the declaration of a recruiting officer that he was eighteen years old. During his service he distinguished himself for neatness, bravery, and an incorruptible discharge of duty.

On a call by the general of his division for the best man in the regiment, he was detailed to take a bounty-jumper from Indianapolis to Baltimore. Being offered a commission as first lieutenant, he had to decline it because he could neither read nor write. At the end of the war he was mustered out of the service as color-guard, a slight mark of honor forced upon him by his superiors, and one from which his want of learning did not bar him.

*

**SILAS.**

*

Short of stature, with broad shoulders, thick neck, and solid frame, Silas was a marvel of strength and as agile as a cat. The writer has seen him take up and carry on his shoulder a log, under which two ordinary men would stagger. Neatness and cleanliness were two of his greatest virtues, in which many a high-born white man might have deemed it an honor to be his peer. For eight years or more he had lived at Oldtown,[2] Maine, and had acquired the reputation of being the quickest and most daring log driver on the Penobscot, and his services were always in demand and brought the highest wages. For five years he had served the writer as guide in the Maine woods, and a

---

[2] 2020: Commonly known as, Old Town.

more devoted and thoughtful servant and friend would be hard to find. Entering into the spirit of exploration which prompted the writer's forest tours, he often devised ways to overcome obstacles, and pushed forward where others would have faltered or turned back. His only weakness was that curse of the white man, strong drink, a few swallows of which were enough to set his brain on fire and make him quarrelsome and vindictive, characteristics which at other times seemed to form no part of his nature. Poor fellow! He came to an untimely end in the spring of 1882, while at work in the woods of Northern New Hampshire. He fell on his head from a lodged spruce tree, up whose slanting trunk he had climbed to cut away the interlocking branches. Death was instantaneous.

Farewell, Silas! May thy life in the happy hunting-grounds be peaceful! May atonement for mortal weaknesses be tempered and sweetened by memories of kind deeds done to others here below, of duties faithfully performed, of an unswerving honesty of purpose and an unflinching integrity!

As our craft passed beyond Trout Point and the Three Sisters and approached the face of Kineo, the afternoon sun cast many a dark shadow on its rocky side before us, and the screams of a pair of cliff-eagles, that hovered over its summit, recalled to us the old legend of the squaw who once lived in a hole up there, and, thrusting out her head, laughed hideously at canoemen as they paddled by.

The three hours of our passage were spent in forming a better acquaintance with our guides, neither of whom had Captain Sartor ever seen before that day; and in laying out and discussing our route, with which not one of us was entirely familiar. The shores of Moosehead Lake in North Bay are less broken than elsewhere, and the few mountains on its eastern side have but a momentary interest for the passer-by. We had a fine view of Mount Ktaadn as we approached the head of the lake, and at five o'clock landed at the end of the long pier which connects deep water with the shore. Here, at the cosey (*cozy*) little hotel kept by Mr. Simeon Savage, we passed a quiet night, the last we should spend indoors for several weeks.

The great North Bay of Moosehead Lake separates into two arms, at the heads of which are what are called respectively the Northeast Carry and the Northwest Carry, roads which lead to the West Branch of the Penobscot, some two miles to the north. Before the use of modern conveyances was known in these wilds, the Indians carried their canoes on their backs over the portages, which then were but narrow paths through the dense forest. In later years the enterprise of the logger built a tram way across the Northeast Carry, over which supplies were drawn by oxen, to be distributed among the "logging camps" on the Penobscot. Theodore Winthrop has immortalized this road and its quaint and patient motive power, which linked civilization with the world beyond.[3] But the tramway is no more. A destructive fire, which laid waste many acres of forest growth, ruined it as well. Its vestiges, in the form of a few charred and decaying timbers in the rank shrubbery at the side of the present wagon-road, are hardly conspicuous enough to cause more than passing comment. However, "Ill blows the

---

[3] "Life in the Open Air."

wind that profits nobody." This burnt district of late years has yielded a large supply of delicious blueberries.

The land between Moosehead Lake and the Penobscot rises gradually from each of these waters to a height of perhaps fifty feet or more above the former, the river opposite the Northwest Carry being 11.36 feet higher, and opposite the Northeast Carry some feet lower, than the lake.[4] The easterly road, by constant use, has been worn down on the southern slope to the underlying gravel and rock. Its bed is a foot or two below the level of the surrounding soil, and, like all carries of the kind, it offers incomparable inducements to serve as drainage for the water that collects on both sides. The tourist can therefore depend upon having a wet walk over part of the carry at least, except in very dry seasons.[5]

Our canoes and camp-equipage having been carefully loaded on the wagon, the impedimenta in the wagon-bed and the canoes inverted over it and fastened side by side to two cross-sticks, Captain Sartor and the writer bade Mr. Savage good by, and, leaving the guides to follow with the load, walked ahead. The exercise in the keen morning air soon put in a glow our faculties both of mind and body. The bright sun in the eastern sky shed a lustre over our path that reached far beyond the bounds of physical sight, and the old enthusiasm at being

---

[4] In Wells's "Water Power of Maine," p. 101, the Penobscot is said to fall 34 feet between the two "carries," which would make the difference of level between lake and river at the Northeast Carry 22.64 feet. This is commonly thought to be too great.

[5] **2020**: The Northeast Carry is now known as the start of the Penobscot River Corridor. The connection is from the Northern Forest Canoe Trail across the North Bay of Moosehead Lake.

again about to enter Nature's wild domain fired the writer anew. Who can describe the sweetness of that first whiff of forest aroma! The drying branches of some prostrate fir-tree load the air with a fragrance one would fain drink in in never-ending draughts. Our old friends, the birches, nod a joyous welcome, as they rustle in the rising breeze. The bushes, berries, wild-flowers, mosses and lichens, all revive some pleasant memory. Our pulses throb with new life, our step grows elastic, and we are already creatures of a different mould from yesterday.

The Northeast and Northwest Carries are of about an equal length, and James Russell Lowell, when he crossed the latter with a load on his back, estimated the distance at eighteen thousand six hundred and seventy-four miles and three quarters.[6] More sober measurements, however, make the length of the Northeast Carry two miles and forty rods.[7] At its northern end, on the right bank of the Penobscot, is the farm of Joseph Morris, the last human habitation we should see for nearly twenty miles.

Embarking in our canoes, we were soon gliding over the glassy surface of the West Branch, as this river is familiarly called, towards the mouth of Lobster Stream, which joins it about two miles and a half below Morris's. From this point it is a like distance, through equally quiet water, up into Lobster Lake. At the mouth of Lobster Stream there is an island, on one side of which the flow of water from the lake is almost directly against the course of the current in the Penobscot.

---

[6] "A Moosehead Journal," in "Fireside Travels."

[7] **2020**: This distance from Lowell was tongue-in-cheek. See excerpt.

| A MOOSEHEAD JOURNAL | **2020 Edition:** |
|---|---|
| choly, and which was finished, and the rest of his voice apparently jerked out of him in one sharp falsetto note, by his tripping over the root of a tree. We paddled a short distance up a brook which came into the lake smoothly through a little meadow not far off. We soon reached the Northwest Carry, and our guide, pointing through the woods, said: "That 's the Cannydy road. You can travel that clearn to Kebeck, a hunderd an' twenty mile," — a privilege of which I respectfully declined to avail myself. The offer, however, remains open to the public. The Carry is called two miles; but this is the estimate of somebody who had nothing to lug. I had a headache and all my baggage, which, with a traveller's instinct, I had brought with me. (P. S. — I did not even take the keys out of my pocket, and both my bags were wet through before I came back.) *My* estimate of the distance is eighteen thousand six hundred and seventy-four miles and three quarters, — the fraction being the part left to be travelled after one of my companions most kindly insisted on relieving me of my heaviest bag. I know very well that the ancient Roman soldiers used to carry sixty pounds' weight, and all that; but I am not, and never shall be, an ancient Roman soldier, — no, | The passage from **Lowell on his 1853 Moosehead trip**.[6] James Russell Lowell (1819-1891) was an American Romantic poet, critic, editor, and diplomat. He is associated with the Fireside Poets, a group of New England writers. |
| *Public Domain Copyright.* | |

The lake is fed by several short and small brooks, which ordinarily are quite inadequate to keep its level above that of the river. Upon any rise of the latter, its waters flow easily, and as a matter of course, into the lake, sometimes raising the water level there eight or ten feet in a few days. From this circumstance, doubtless, the lake takes its Indian name, Pes'kébégat, "Branch of a Dead-water."[8]

[8] Another interpretation of this word is "Branching Lake," or "Split Lake," which may be a better one than that given in the text. The form of the lake well corresponds to it.

An amusing incident is related of a logger, whose "drive of logs," during his first experience on the West Branch, went up into Lobster Lake. He supposed all the while that he was following the main river, until he reached the lake, when he was completely lost, and thought himself bewitched. However improbable it may seem that this man and his "crew" should have been unacquainted with the two streams at this point, the story furnishes a very good illustration of the characteristics of the place.

On Lobster Lake we loitered for a day in well-spent idleness. Our camp was on a point of land from which we took in the receding sweep of the Lobster Mountain range, the Spencer peaks, long Sabōta'wan and rounded Kōkad'jo, and far-off Ktaadn with its lovely morning shadows. Steep rocky promontories crowned with Norway pines[9] jutted out on our right into the lake, while between them lay most seductive reaches of sandy beach. A picturesque island was on our left, and far beyond it stretched West Cove, which, with the lake and point of land between the two, helps to form the fancied outline of a lobster's claw.

Captain Sartor with his brush, and the writer with his camera, passed a portion of the day together, while the guides opened our boxes and arranged their contents in convenient packages. A visit in the afternoon to Little Lobster Lake, the

[9] **2020**: The red, or Norway pine (Pinus resinosa), is native to North America. It is the state tree of Minnesota. The name may have been given by Scandinavian immigrants who likened the area forests to those of their home country. Norway, Maine is also an area rich with red pine.

bagging of three ruffed grouse, and a quiet "paddle" towards evening, completed the experience of our first day in camp.

After supper, while the guides were chatting and smoking their pipes, and Joe was posing for a picture, the writer wandered away from camp down to the moonlit shores. The day had been uncommonly fine; the night was superb. The forest lay in calm repose. Its silent aisles were curtained with darkness palpable, save where the moon's bright shafts, entering here and there the lighter foliage, and struggling through its network of leaves and piny needles, were sifted into a softer, mellower light. The outlines of lofty spruce trunks, like sentinels at their posts, were just visible at the edge of the near obscurity, and up the sides of the spectral birch trees by the shore trembled a gentle shimmer of light, reflected from the rippling waters of the cove. Not a breath of air was stirring. The ripples came, as if impelled by an unseen hand from some distant source, to lap the warm and sandy beach, and there dissolved without a murmur.

A dread stillness prevailed, a silence that could be born only of night, weird and supernatural. And yet in this very silence of Nature in her gentler moods there is a rhythm, as it were, that, acting on the spirit, charms and soothes, — a chord whose each successive vibration, emanating from some wondrous hidden source, acts with varying force upon our rougher nature, harmonizes at each pulsation its rude components, and moulds us insensibly into better, truer beings. Hardened indeed must he be over whom these charms have no refining power. The beautiful waters, woods, and hills, the subtile influences that hover over them, the expanding freedom of body and soul, as they reach forward with ecstatic longing as if to kiss sweet Nature's self, — these are for him who seeks seclusion from the world in the wild-woods' vast domain. There is fit place for rest and meditation; a fitting altar on which to sacrifice our baser thoughts and passions, long endeared to us; a place from which, with strength of body, mind, and will renewed, we may reissue well prepared for sterner duties.

*

**READY FOR THE START**

*

# Chapter III

*

Penobscot Valley. — Rapids. — First Morning in Camp. — Breaking Camp.—"Pitching" and Loading Canoe —A Caribou. — Measuring Distances. — Fox Hole. — A Mess of Trout . — A Mishap. — Reflections. — Running Rapids. — A Fifth Passenger. — Squirrels.

*

**BETWEEN** Pes'kebégat[1] and Chesuncook the Penobscot, in fifteen or sixteen miles, falls a hundred feet or more. Over about half of this distance the water is sluggish; over the other half, it has a noticeable current, which at times increases to the velocity of a rapid, notably at Big Island Rips, Rocky Rips, and Pine Stream Falls. The first two of these rapids can generally be passed over with ease, even by a heavily laden birch canoe; but at the last-named, at least one occupant of so frail a craft will sometimes find it prudent to disembark and walk around the upper and more difficult part of the falls. The river has a width of from thirty to seventy yards, or even more when groups of rich alluvial islands divide its course. Most of these are crowned with a rank growth of grass, and on them here and there a stunted elm-tree shows its outlines against the vacant sky.

The valley of the Penobscot here is quite broad. The river's banks are from six to eight feet high, and it is not unusual in spring freshets to see the water even with their tops, or

---

[1] **2020**: Lobster Lake

overflowing them. Beyond the banks the country runs off into wide level sweeps of rich land, covered with a mingled growth of fir and hardwoods, or here and there gently rises in undulating swells above the intervale. The river's even flow is seldom broken by a rocky ledge, and the serried lines of trees on both sides shut out from view the distant hills and mountains, save where they give a few glimpses of the Spencer peaks, or a glimpse or two of Mount Ktaadn. Now and then the light breaks through a grove of poplars, the after-growth of some forest fire, and beyond them may be found a mossy bog, or the grassy banks of some sluggish brook.

Over this course lay our second day's journey. The guides were up by daylight, and the sputtering of their new-made fire soon warned us that we too should rise. A generous application of the waters of frost-nipped Pes'kebégat soon opened wide our heavy eyelids; for the first night in camp does not always bring sleep to the drowsy nor rest to the weary, and, although the writer had no special complaints to make, there were signs about the Captain which made it politic not to pass the compliments of the morning too confidently, nor to discuss with too much enthusiasm the pleasures of camping. A neutral tone was therefore given to the conversation, and although the writer noticed the Captain's eye fixed on him in a very meaning way during one or two of the pauses, the dangerous period— until breakfast-time — was successfully bridged over, and safety was then assured for the day.

After a breakfast of bread, bacon, and boiled potatoes, the "three Bs," as the Captain, from its frequent recurrence, soon learned facetiously to dub our bill of fare, — we fell busily to work to break camp. To each active man was assigned his routine duties. The writer's part was to fold up the blankets, and to put in order our bag of knicknacks and extra clothing, which in sundry parcels, together with hats, boots, and hardware, had served for "heading" during the night. These, with gun, camera, and box of plates, constituted all his cares, while the Captain, being a guest, and more or less aesthetic, stood aloof, and viewed the scene with an artist's eye.

Every guide makes or should make a practice of over hauling his birch canoe each morning before putting it into the water, and when an eye in the bark has been freshly broken, or the "pitch" rubbed off, a new application of that substance must be made. The damaged "eyes," being often very minute, are not always easy to find. Where they cannot be seen at a glance, the guide applies his lips to the suspected surface, and by a sucking process can detect a leak at once. The different spots, as soon as found, are marked with a pencil, or with a charred stick from the camp-fire. The broken eyes are then cut into on either side along the fracture, and at an angle of forty-five degrees to its plane, and thoroughly dried, either with a fire-brand, or by being exposed to the sun. These preliminaries over, the guide takes his pitch dipper from the coals, where it had been previously placed, and applies the melted resin to the bark, smoothing and pressing it into place with his moistened thumb. Incisions made carefully in this way hold the pitch a long time, as the latter cannot readily be rubbed off enough to reopen the break. There are often leaks where the bark has been cut or torn by contact with the rocks, and each new scrape, however slight, is apt to break the brittle resin and reopen the wound. The canoeman is sometimes compelled, for this reason, to stop in the middle of the day, unload his birch, and as it were go into dry dock for repairs.

*

**PITCHING THE CANOE.**

*

To load a canoe properly, when about to take a journey in it, requires time and patience, with a good deal of experimenting. First of all, the bulk should be so arranged that the man in the stern may have perfect freedom to move about when he stands up to guide the craft. The bow, too, must be so weighted that in rapid water it shall feel the effect of the current and quickly yield to it. Of the first importance is it, however, that the canoe should be trim. Where there are two canoes in a party, after the division and arrangement of luggage has been once established, the operation of loading becomes mechanical.

On the morning in question, owing to many delays, it was nine o'clock and after when we left Lobster Lake, and found ourselves sailing down the outflowing stream before a brisk southeast wind. Suddenly, on turning a bend, we saw a caribou[2] (Rangifer caribou)[3] approaching on the opposite bank. Our sense of sight was scarcely quicker than his sense of smell, for although two hundred and fifty yards away he made two or three sniffing movements of the head towards each side, and, turning into the alder-bushes, disappeared before we could recover from our surprise. "That caribou must have his cane with him, he leave the country so quick," exclaimed Joe, whose regret at the absence of fresh steak from our table manifested itself for several days afterward.

This incident created quite an excitement in the Captain's mind, and although we cautioned him that several days must elapse before we could lawfully take game, he solemnly drew forth the sole weapon he possessed, a small thirty-two caliber revolver, impressively laid it before him on the bottom of his

---

[2] **2020**: Caribou, once plentiful in Maine are a rarity. Due to over-hunting, it has been illegal to shoot caribou in Maine since 1899. There have been two efforts to restore caribou to the Maine north woods. The first in the 1960s, and a second in the mid-1980s. The effort was abandoned due to complications.

[3] **2020**: Rangifer tarandus caribou is known as the woodland caribou, boreal forest caribou and forest-dwelling caribou, and they are a North American subspecies of the reindeer.

canoe, and vowed that no moose or caribou should come near him, law or no law,— not near enough to "bite" him, anyhow.[4]

The average sportsman in this part of the world, especially when in a canoe, is apt to be somewhat careless in not having his gun within reach, and disengaged from everything that might prevent its immediate use. Nothing will sooner make him realize the importance of being ever on the *qui vive*[5] than the sight of game, and we were no exception to the rule. The guides' conversation dropped to monosyllables uttered in an undertone, while the Captain and the writer devoted their attention exclusively to scanning the banks of the stream as we glided onwards and into the Penobscot. Several years before this time a friend of the writer, unarmed of course, had suddenly come upon a large member of the cat tribe that was swimming across this very stream, and we knew that bears had not infrequently been seen on the shores of Pes′kebégat.

The West Branch, in the parlance of loggers, "had on a good pitch," and for two and a half miles after we first turned into the river we found that the current hurried us along with considerable speed. In the channels the water was, on an average, over two feet deep, and the bed of the stream was for the most part comparatively smooth. Rooted in it were patches of a coarse grass, whose blades were flat and narrow, like ribbons. The long snake-like forms writhed and bent in the

---

[4] **2020**: In 1880 the Maine Legislature assigned two Maine Fisheries Commissioners to enforce the laws related to important game species. This was the beginning of what is now the Department of Inland Fisheries and Wildlife.

[5] **2020**: French term to mean a state of alertness or watchfulness.

current, their heads now sinking low beneath the water's surface, and again rising as if to take breath.

It has been the writer's experience, that in still water, when there is little or no wind blowing, two men paddling steadily and with moderate force can impel a lightly laden canoe at the rate of a mile in about seventeen minutes. The speed of a canoe is rarely accelerated, on Maine rivers at least, to the rate of a mile in eight minutes, and then only for short distances and over reaches of comparatively unobstructed swift water. The method of estimating distances on the water by thus timing one's progress is often of great service to the tourist, and helps to make him independent by fixing his locality approximately, and aiding to determine when or where he shall stop for the day.

From the mouth of Lobster Stream, it took us two hours to go eight miles, as far as Sears's Clearing, a deserted spot, where, ten years ago, a squatter let daylight into a small patch of ground, and sought to raise sustenance for his growing family. Then shortly we reached the mouth of Ragmuff Stream,[6,7] which but for its noise and clatter we might have passed by unnoticed.

The guides now began to talk about dinner; but we decided that, as our fishing days were numbered, we would go on to the Fox Hole, less than two miles further, and, while dinner was preparing, cast our flies into that noted trout-pool. It took us half an hour to reach Big Island, and, bowling down the rapids

---

[6] P'tay-week-took, or Pay-tay-week-took; Burnt-Ground Stream.

[7] **2020**: Ragmuff Stream is the first of the two Ragmuffs. "The Maine Atlas and Gazatteer," depicts *Little Ragmuff Stream* at another five or so miles north of this point. Maine Penobscot River Corridor map includes excellent detail on the locations and camp site.

on its eastern side, we soon turned sharply to the left into a narrow channel within reach of the bank, and checked our course at the mouth of a little inlet forty or fifty feet long and perhaps half as wide. Several logs, drawn in here by the eddying currents of high water in the spring of the year, had lodged in a confused mass near its upper end, and altogether the spot was one in which ninety-nine hundredths of the passers-by would never suspect the presence of a trout.

Cold subaqueous springs are thought to exist here, and if they do, they furnish a refreshing retreat to the fish during the warm summer months. The trout, however, is an uncertain creature, and a trout-pool today may be an empty pool tomorrow. As the Captain put his rod together, and fastened to his leader a Montreal, a brown-hackle, and a red-ibis[8] fly, Joe quietly said to him, "Mebbe you catch plenty; mebbe you don't catch any. We'll see."

Sartor made his first cast into the middle of the pool. No response from below. A second and a third in the same place were equally fruitless. On the fourth cast the tail-fly, the Montreal, landed at the side of one of the logs and in its shadow. A swirl, a flash from shadow into sunshine, a splash, a taut line and bending rod, all followed in such quick succession that what before seemed an improbability was now a certainty. There were fish in the pool. When brought to net the first catch was found to consist of two trout of a pound weight each, on the tail and middle flies respectively.

---

[8] **2020**: Thought to be related to the 'Scarlet Ibis' from England, developed early in the 1800s.

Sartor repeated his casts for ten minutes or more with varying success, while the guides, who had already built a fire from pieces of dry driftwood which lay scattered about in profusion, busied themselves with cleaning and frying the fish and preparing the other courses of our meal. Determined to take "just one more," the Captain made a vigorous effort, and landed his tail-fly, not in the water, but on a spruce log, in whose bark the hook was soon imbedded, while the remaining flies dangled above the water in a manner too enticing to be borne by any fishless stolid than the sucker. The result was, that in a twinkling the hand-fly was gracefully taken by a trout whose whole length showed out of water, the leader snapped from the added strain, and instantly went out of sight, together with the now disengaged tail-fly.

"Well, now, that's rather hard on a fellow," complained the Captain. "Forty cents' worth of leader and about fifty cents' worth of flies[9] for one pound of trout, — and didn't get the trout either. They say it costs a visitor to Maine five dollars a pound, on the average, for all the trout he takes. It might be interesting to know what he pays for what he *doesn't* take; but I fancy statistics on the latter subject would not be forthcoming as readily as in the former. As for that matter, though, what's the use, anyhow, of catching a big trout or shooting a moose, unless you can brag about it? Why doesn't the legislature change the game laws, and let people shoot in September? They do so now, in spite of the law; and have to burden their consciences still further, by denial or by deceit, in their efforts to escape

---

[9] **2020:** The current price for a single Dark Montreal fly can run upwards of over $4.00.

detection and punishment. You don't suppose I'd give a fig to kill big game, do you, if I couldn't tell my friends all about it? Why then, I say, can't we shoot game in September? Then a -"

"Simply, my dear fellow," interrupted the person addressed, "because the country is already full of men, whose fish-stories are as stale as would be the poor subjects on which they are built, if the latter's carcasses had been given a like airing. The responsibility of letting loose upon the community a kindred set of game story-tellers is more than any sane legislator cares to take, and this is precisely what would happen if you and others of your ilk could lawfully shoot big game in September."

The trout were now ready, and four hungry men seated themselves as comfortably as they could, in near proximity to the frying-pan. To the Captain, as hero of the day, was awarded the use of our only bucket, which on this occasion served him as table. Aside from its awkwardness on a carry, the bucket or firkin is a very useful and desirable appendage of a camper's outfit. It holds the dishes and a host of smaller articles which might otherwise become lost or misplaced. It serves the purpose equally well of table and stool, while its cover is a meat-board or waiter at the cook's pleasure. The others of us, on this occasion, contented ourselves with that less reliable substitute for a table, the lap, but managed all the same to do ample justice to the Captain's trout.

After dinner and a quiet smoke, we proceeded leisurely down the river. There was no need of haste, because we proposed to camp near Chesuncook Lake, and had less than six miles to go. From the head of Rocky Rips, as we shot down into Pine Stream dead-water, we had a good but momentary view of

Mount Ktaadn, and soon passed the mouth of Pine Stream,[10] whose dark water was rendered darker still by the shadows of the spruces on its banks.

*

**LUNCH BY THE WAY.**

*

At Pine Stream Falls the river pitches over several broken ledges with much impetuosity, and a clear head and quick arm are required on the part of the canoeman to keep his craft off the rocks and out of the boiling waves below them. Ordinarily, when the bow is occupied by a person not a thorough canoeman, the entire responsibility of managing the canoe rests upon the man in the stern. He stands erect, one foot in advance of the other, and his body turned partly towards one side. With his long-reaching setting-pole he controls his birch's movements, and contrives by a series of checks or "snubs,"

---

[10] 'Mkaza-ook-took; Black Stream.

rapidly made on alternate sides, to let her glide slowly down. Now and then with a vigorous push from behind he makes her shoot along towards some better channel, and avoid a sunken rock, whose ripple shows itself a foot or more lower down the stream. Again, finding the water shoal, he backs "quartering" across the stream; for he sees the water in the river-bed set that way, and knows that yonder he will find the better channel. The bow-man sits with ready paddle to help attain some point of vantage, or fend the bow from some projecting rock. If a tyro, he often shudders lest his craft strike a boulder which seems directly in its course, but his guide knows well that the water sets off from that apparent obstacle, and the current's force is felt just as the canoe seems about to strike. The birch barely escapes, and as it glides by in safety the bow-man breathes again. Sometimes, however, the iron point, or "pick," of the setting-pole slips over a smooth ledge, or is caught between two rocks, and the pole jerked from the holder's hands. The canoe is quickly at the mercy of the rocks and water, and escape is possible only by some lucky chance. To jump overboard is one's first impulse, but that is rarely practicable and very often dangerous. Whether this is done or not, the canoe, hurried on by the swift waters, strikes a sunken rock, bow first, quickly swings around against another, and tips its load pell-mell beyond it into the seething rapids, itself perchance a shattered wreck. Its luckless occupants, if haply they be uninjured, must scramble to the shore, and they may be sincerely grateful for whatever of their camp stores may be saved and fit for use.

Two men accustomed to handle a canoe can take it with ease up or down an ordinary rapid. Each has a pole, and stands erect.

The bow-man now selects the course, and keeps the canoe directed in it. On him rests principally the responsibility for their safety, while the stern-man's function is machine-like and his movements are the counterpart of his companion's, except that when going up over a difficult "pitch" they become supplementary. The one then holds the canoe in a position once gained, while the other plants his pole anew for another push.

*

**COMING DOWN THE RAPIDS**

*

The act of running rapids in a canoe is always exhilarating. To a person of good nerves who tries it for the first time, it is apt to be nothing but pleasurable; but one who knows its dangers never enters upon it without some slight fear or trepidation. And yet, the danger passed, one is ever ready to face it again — with a skillful steersman.

We passed Pine Stream Falls without mishap other than the shipment of a little water, and after half an hour's paddling reached two large piers in the river, that help sustain the boom when it confines the spring "drive" of newly cut logs. Just beyond the piers we came upon a chipmunk (*Tamias lysteri*)[11] swimming across the stream. The little fellow seemed quite tired, and when the writer's paddle was held out to him and he felt its firm support beneath him, half drowned he walked up its handle and into the canoe. There he sat perched upon the highest of our bags, drying his dripping jacket, until we neared the river-bank, when he jumped overboard and swam to land.

How intimate one soon becomes with the squirrels in the woods! The curiosity with which from a low limb of some fir-tree they investigate our intrusion into their haunts, the chattering and scolding with which they resent it, accompanying each grunt-like note with a sort of electric twitching of the hind legs, and their scampering around the tree and up its opposite side when we venture to move, just the tip of the little head appearing from time to time on either side of the trunk on their upward course,— these are phases of woods life with which all campers in Maine are doubtless familiar. These busy little creatures are often seen to loiter in their forest gambols, and even come down from their high perches among the spruce cones, to listen to a sound made with the human lips that resembles the squeaking of a musquash. At such times they will sit in perfect silence for many minutes, as if charmed, complacently cocking their little ears, and with one or both fore feet folded against their snow-white breasts. On one such occasion the writer approached a squirrel until he could all but

---

[11] **2020**: Tamias striatus lysteri, the Eastern Chipmunk, is one of eleven subspecies of Tamias striatus. Hubbard had shortened the hierarchy.

touch him with outstretched hand, the little fellow remaining perfectly quiet.

The acquisitiveness of the squirrel sometimes makes him a nuisance even in the woods. An incident was once related to the writer of two campers who, on the morning of the day on which they were to go home, left in their tent the only remaining food they had, and which they intended to eat for lunch. This consisted of some large home-made soda biscuits wrapped securely in a piece of strong paper. The campers came back from their morning excursion about noon, tired and hungry of course, and searched everywhere for their lunch. Not a biscuit was to be found, nor even a crumb of one. By a lucky chance, however, one of the men had a dry crust in his pocket, and when he and his companion sat down to eat this outside of the tent, high up on the limb of a neighboring tree they spied a red squirrel moving along with one of the missing biscuits in his mouth!

From the piers we had a fine view of Ktaadn, and soon after passing the cottage of Ansel Smith, one of the Chesuncook pioneers, we landed on the right bank of the stream and made preparations to camp for the night.

# Chapter IV

*

Preparations for Night. — The Captain's Opinion on Camping. — A Misty Morning. — Chesuncook. — Up the Umbazookskus. — Smith's "Jumper." — Its Effect on Moosehead Guides. — Making a Portage. — Mud Pond Carry. — Native Modesty.

*

**THE** choice of one's campground is a matter of no little importance to the sojourner in the woods, and often requires the examination of several sites before all the requisites, or the principal ones at least, of a good site are found combined. This fact, and the other one, equally patent, that no man can see well enough after dark to cut or gather wood sufficient for the camp-fire all night, seem to escape the perception of the average camper, often much to the dissatisfaction of his guide. We made it a rule, in our party, to go into camp at or before five o'clock, an hour quite late enough for October days, and yet one which generally gave the guides ample time to pitch the tents, cut their wood, build their fire, and prepare for supper, before the shades of night had enveloped us. The site chosen for our second camp did not possess all the requisites of a first-rate campground. In the first place there was very little wood near it, either green or dry, and secondly, we had to scale a steep and muddy bank to reach it. The ground was level, to be sure, but hard and unyielding, and no boughs could be had for our beds. We soon, however, found a substitute for boughs in the long, rank grass which grew around us; and Joe, in lieu of a scythe, wielded an immense huntingknife which he wore at his belt, and in a few moments had cut enough grass for both tents. That knife, whose blade was twelve inches long, in the outset seemed to us a monstrosity, and excited no little merriment from its uncouthness, but before the end of the trip we thought better of it. It was penknife, butcher-knife, scythe, or axe, as occasion required, and did its work well in each capacity.

Our tents were soon up, "bedded down," and all the luggage from the canoes stowed away under them, except such as was wanted immediately by the cook. A cheerful fire sputtered and crackled between the tents, whose openings faced each other; and while Joe began to prepare the evening meal, Silas took the two canoes out of the water and turned them over on their sides to drain and dry. The Captain and the writer, on landing, had discovered a large quantity of luscious-looking high-bush cranberries, which hung in enormous red clusters from their bending stems. We picked a pailful, and, with sugar added, soon reduced them over the fire to a delicious sauce. The Captain during the latter operation had quietly withdrawn into our tent, where he was discovered on hands and knees, as if searching for something among the blankets.

"Looking for those quarters you lost out of your pocket last night, Captain?"

"No, I'm not, but I'm going to give these quarters a pretty good overhauling though, before I turn into them, and I propose to have, among other things, a better pillow than I had last night. I awoke two or three times — I was going to say; but come to think of it, that was my normal condition. At any rate,

I realized several times that I had doubled up my ear and made a pillow of that, — no pleasant sensation, I can assure you, — and it took me some time to get the creases out too. Talk about your soft, spicy beds of balsam boughs! I slept on bare slats, and I know how many, for I counted them at least a hundred times. Then there was a root right under my fifth rib when I lay down, which at first seemed only a trifle, but that root grew in half an hour to the size of a log. If I turned over on my other side, the ground under my hip was higher than that under my ribs, and I was soon ready to collapse. Then I tried lying on my back, but an industrious spider was soon busy connecting my nose with the ridgepole of the tent. After a while I managed to get both eyes shut, and was sinking into a doze when — mother of Moses! — I thought a wild-cat was scaling the tent on the outside. What a fearful noise! Then another wild-cat ran up after the first, and the first went down the opposite side. I kicked you, but you wouldn't wake, — you and the guides were too busy snoring. Heavens! I would rather take my chances camping on Boston Common without a tent. When finally, I did go to sleep, it seemed but a few moments before my feet and right shoulder were freezing cold, and I found them uncovered. From that time until after daylight my constant prayer was that the guides might soon get up and build the fire. Where, I should like to know, are all the comforts and enjoyments you held out so alluringly before me when you inveigled me from home? I can rough it well enough in the daytime, but when it comes to the night, I must say, I'm a little particular, and don't want my bed on the side of a hill or on a woodpile. By the way, Joe, what were those creatures that made such a racket on the tent? Weasels?"

"They must be — mice, I think," replied Joe, "little fellows with these little short tails. They lookin' for somethin' to eat."

Silas, who had heard the Captain's recital of his first night's experience in camp, and had been much amused by it, now carefully examined our beds and freed them from every substance that would be likely to obtrude unpleasantly before morning, and ever afterwards this was one of the attentions he regularly showed us. We noticed, however, that often, when he seemed on the point of doing us some unusual favor, Joe would apparently restrain him, and Indian etiquette made the younger defer to the older guide. As their conversations were always in a tongue strange to us, we could only infer their thoughts from their acts, and our inferences were still further strengthened by the absence on Silas's part of many attentions it had been his wont in previous years to show.

The next morning the writer was awakened by an exclamation from Sartor, who bade him rise and look out upon the scene. Sartor was already busy transferring it to canvas. The sleeping river at our feet was as smooth as a mirror, and the air was filled with a light, impalpable mist, which gave the landscape a wide-reaching and dreamy perspective, toned down its colors, and tinged them with gray. Three solitary elms drooped motionless over the opposite bank, whose low and bushy surface extended from us in a point into the placid lake beyond.

Physically considered, Chesuncook Lake is one of the most interesting lakes on the course of the Penobscot. Besides the latter stream, two others of considerable importance and several smaller ones flow into it, which drain a large area and contribute not a little to the water-power of the out flowing Penobscot. It is the reservoir in which these various forces gathered before their combined assault upon the barriers which once separated them from the ocean. With the exception of the views it offers of Mount Ktaadn, the scenic attractions of Chesuncook are not of a high order. It presents an almost unbroken line of forest except at the northern end, where for half a century a few sturdy settlers have been struggling with the wilderness, and have succeeded so far in overcoming it as to make several thrifty farms.[1]

We broke camp at nine o'clock, and paddled quietly through the dissolving mist out of the river into the lake, and across the latter into the Caucomgomoc.[2] Just as we started, a canoe passed us in which were two men, one of whom in answer to our inquiries said that Ansel Smith — Ansel the younger — was still on Mud Pond Carry with his horses, and expected to haul us over. After they had gone, Joe said the speaker was Charley Smith, Ansel's brother.

An eighth of a mile or less from its mouth, the Caucomgomoc divides, and the right or eastern branch is known as the Umbazookskus.[3] In ordinary summer seasons, and when the gates of the Chesuncook dam are up, the

---

[1] **2020**: Chesuncook Village was settled in 1849 as a logging community. The 150-year old historic Chesuncook Lake House Inn, built in 1864, was destroyed in a fire in March of 2018. The re-build began in 2019.

[2] Or Kahkoguamock, Big-Gull Lake.

[3] "Meadow Place."

Umbazookskus for four or five miles from its mouth might be said to wriggle through the meadows which line it on both sides, if it only had life enough to wriggle. It may have wriggled once, long ago, but went to sleep in the act, and has never waked to stretch out its lazy form. Fortunately for us the Chesuncook gates were down and the meadows were temporarily flooded, so that we took a straight course across them up the stream. As we had been forewarned to do so by Charley Smith, we fired several shots on our way up the meadows, and heard an answering shot from their head.

**UP THE UMBAZOOKSKUS**

*

As we came opposite Big Brook, a tributary on the right, our course changed decidedly to the left, and the channel seemed to lose itself among the rank water vegetation, so that for a few moments we were in doubt where it lay. Finally, it reappeared, now quite distinguishable from the flooded growth of stunted bushes at its side. The open space between the lines of forest on each hand now rapidly narrowed, a strong current manifested itself in the stream, and we found ourselves on a small brook opposite a well-worn landing-place. Smith's "jumper," a rough but substantial-looking two-horse sled, stood a few yards from the water's edge, and we had the choice either of being hauled over to Mud Pond directly from here, or of going up the stream and across Umbazookskus Lake to the beginning of Mud Pond Carry. In the latter case the expense would be less, but we should lose time, as Smith must bring his horses to where the jumper now stood, and then take the latter around to Umbazookskus Lake, to accomplish which he would have to come back some distance over the carry. As the morning fog had turned into a cold drizzle, we adopted the former plan, and while lunch was preparing Sartor and the writer jumped into a canoe and paddled downstream again to the edge of the opening. Here we built a small fire, by the side of which, on his stool and under his wide-spread white umbrella, Sartor proceeded to "take down" the dim and misty scene before him. A wide waste, at the lower end of which the converging lines of forest almost met; a flat, frizzled surface of rich dark-brown, changing to gray where the autumn frosts had stripped it of its foliage; through it projecting at intervals some gaunt and shapeless tree-stem, moss-bedecked; in the foreground a bit of water dotted with lily-pads and lined with rushes; — all this, enveloped in a misty veil and canopied with dull leaden cloud-masses, was the Umbazookskus. No sound was there to break the solemn stillness, save the rusty clatter of the great blue heron (*Ardea herodias*), as he hastened over us on breeze-creating wing.

On our return to the landing we found that Ansel had arrived with his horses. He said he had a little camp on the Umbazookskus end of Mud Pond Carry, had spent the past two months there, and had had plenty of business. He had made that season, as we afterwards heard, something like a hundred and twenty dollars, ready cash; a very respectable sum for a Chesuncook farmer, and in fact enough to make him comfortable for the winter.[4] He had heard our shots, and had left his camp at once to meet us.

In former years, before the appearance of Smith at Mud Pond Carry, comparatively few camping-parties went over that way to Eagle Lake.[5] They went rather up the Caucomgomoc, a route which involved two or three very short carries and a two-mile stretch of hard poling, or they went to Ripogenus,[6] which required but one carry. In those days there "wasn't much to see on Eagle Lake," while Caucomgomoc was a "paradise." The unanimity, however, with which the Moosehead guides changed their opinion when Smith's team appeared at Umbazookskus, was remarkable. Eagle Lake then became the bourne, the Mecca, of all forest pilgrims. According to the guides it was the stamping-ground of the noble moose, — where they congregated night and day. Bears, too, were abundant there. A gentleman had killed three on one brook. In fact, any game that lived in the Maine woods might be expected

---

[4] **2020**: On the order of $2,800 in 2020 at 2.6% inflation.
[5] Pongokwahemook, Woodpecker Place
[6] Not Indian. The Indian name is *Nolangamoik*, Resting Place.

to appear on Eagle Lake at any moment. Marvelous were the tales of its attractions, dressed in all the fantastic coloring that only a second "Old Ellis" or "Uncle Zeb"[7] could paint. The result was that the shores of Eagle Lake nightly gleamed with campfires, and a person hardly dared fire off his rifle there for fear of hitting his neighbor. People who should have known better camped right on the most promising hunting grounds, from which the noise of their chopping and the smell of their camp smoke must have driven off any game that chanced to be near. If, however, Smith did not happen to be on the carry when a party arrived, the guides — "Well, Eagle Lake wasn't much better than some other places; Caucomgomoc was just as good"; — and around they would turn, and paddle back over those five miles of dismal inundation, and seek other hunting grounds easier of access.

It did not take long after lunch to arrange our things compactly in the body of the "jumper," and swing the inverted canoes on ropes, one over the other, over these. The ropes were fastened to stanchions, which ran up from the edge of the sled, and were so arranged that one canoe overlapped the other diagonally without touching it, a precaution absolutely

---

[7] Two well-known characters and hunters of the Moosehead region, the latter of whom still lives. An amusing anecdote is related of Old Ellis, who had received from a gentleman with whom he had been camping ten dollars, with the request to send the latter a good bear-skin. The next spring, Ellis fell upon the track of a bear, which he followed for two or three days, without coming up with Bruin. With dogged pertinacity he kept on, exclaiming at last, somewhat out of humor, "Go it, old Bruin, go it while ye kin! There ain't a hair on yer back that belongs to ye."

necessary owing to the roughness of the road. Sartor and the writer, encased in rubber boots and coat, walked on ahead. The road lay through the dense forest, and, although wet and rough in places, was quite level, and much better than we had expected to find it. We walked about two miles and a half, and came suddenly into the old carry, six hundred yards from where it leaves Umbazookskus Lake.[8]

Mud Pond Carry! What visions that name calls up before one who has seen the carry and walked over it in its heyday, with a load upon his back! when a well-filled brook adorned each end, stopping to filter through deep miry pools, or lost for a time among the roots and moss which alternated with them. As on the Northeast Carry at Moosehead Lake, but in a much more marked degree, owing to its naturally softer character, the soil at both ends of the portage was cut up and rapidly wore away. The tracks made by passing teams soon became the water-shed of the surrounding land, whose moisture seems perennial. These brooks are likely to be ornaments of the carry forever. The roots and stones were allowed to remain on it for years, partly because they furnished a foothold, howsoever slippery, to keep people out of the mire, but principally because no one had public spirit enough to remove them. Since Smith's

---

[8] From observations made by the writer with an aneroid barometer, Umbazookskus Lake appears to be 67 feet higher than Chesuncook, and 25 feet lower than Mud Pond. The divide between Mud Pond and Umbazookskus Lake rises to a height of 55 feet above Umbazookskus and 30 feet above Mud Pond. These figures are considerably above those heretofore current, which make the divide between Mud Pond and Umbazookskus only 72 feet above Chesuncook. See Loring's Report, quoted in Greenleaf's "Survey of the State of Maine," page 63. The carry is 3,150 yards long, or 370 yards short of two miles.

advent, however, and the increased travel that way, some improvements have been indispensable. The road has been "bushed out" in places, so as to admit more sunshine, obstructions have been removed, and the general appearance of things is bettered. The miry pools are generally little more than a foot deep, and under them lies a firm subsoil, so that, with confidence, and — a pair of high rubber boots, one may plunge into them without danger. Indeed, one passage over the carry, with a light load, is by no means irksome; it is the "iteration" that is "damnable."

The novelty of our morning's walk and the primitiveness of our road were sources of no little wonderment to the writer's companion, whose exclamations from time to time at what people will undergo for the sake of a "little fun" were decidedly ironical. His patience, which he had been in a fair way to lose just before we reached the high ground in the middle of the carry, utterly gave out when, wading through the stream which formed the carry's eastern end, he stepped upon a rolling pebble, and almost measured his length in the water.

"I say," demanded he, on recovering his balance, "are there any beavers left in this part of the country?"

"Yes, a few," was the answer.

"Do they ever wander over this way? If so, I wonder they haven't dammed this carry at both ends. I'll venture to say everybody else has whoever walked over it, and thoroughly too. Confusion to it! There isn't money enough in all Boston to entice me into this country again."

The Captain, however, soon regained his composure, and in a few moments, we had reached the shores of Mud Pond, and were resting quietly on a log, awaiting the arrival of the guides.

We were now on the waters of the St. John River, and, by adoption as it were, of the East Branch of the Penobscot, or Wassātegwéwick,[9] as well. Mud Pond, a hallow basin, as its name implies, is little more than a mile wide between the end of the carry and the outflowing brook, and slightly longer than it is wide. Its waters in the course of a mile empty into Chamberlain Lake,[10] the great reservoir of the Allagash-Penobscot system,[11] from which in turn they flow in two directions, towards the northeast and southeast.[12]

About four o'clock the jumper arrived, the canoes and *impedimenta* were quickly taken off, and, as this spot was rather a dismal one to camp on, we prepared to cross the pond to the carry on Mud Pond Brook, where Joe said there was a good dry spot on which we could pitch our tents. In reply to a question from the writer, Smith, catching his breath and averting his

---

[9] The meaning of this word, as given by John Pennowit, is "place where they spearfish" (salmon). The root 'wassā' primarily means "bright" or "sparkling," and hence there are those who say that Wassātegwéwick means "place of the bright or sparkling stream." The secondary meaning of wassā, as given by Râle, is "to take fish by torch light." As this meaning is also given to the above name by the principal hunters among the Penobscots, and as the stream has been noted for its salmon, which the Indians almost always take with a spear by torchlight, the writer has no hesitation in adopting Pennowit's translation as the proper one.

[10] Apmoojēnegamook, or Baamcheenungamook, "Cross Lake."

[11] The name Allagash is taken from Wallagaskicigamook, the Indian name for Churchill Lake, through which it flows.

[12] For an account of Telos Canal, see Appendix V.

eyes, said in a husky voice, almost a whisper, that his charge for hauling our "stuff" over the carry was six dollars.[13] Shades of Charon![14] Had Virgil lived in these latter days, he never could have written, "Facilis descensus Averno,"[15] with such ferry charges as these staring poor mortals in the face at the very outset of their journey. Verily, "O terque quaterque beati....Trojae sub moenibus altis,"[16] &c. Sartor, to whom the writer afterwards turned for sympathy, said that the price was little enough; that he wouldn't have rendered the same service for five times the money.

On the farther side of the pond, as we neared the old dam at its outlet, we met two gentlemen in canoes, with their guides. They were on their way back from a four weeks' excursion, and were camping on the carry near where we afterwards pitched our tents. In the evening they made us a friendly call, and an hour or more was spent in relating to each other our experiences in the woods.

---

[13] **2020**: About $150.00 in 2020.

[14] **2020**: Refers to ferryman of the dead.

[15] **2020**: Facilis descensus Averno Latin for (the) descent to hell is easy. From Virgil's *Aeneid.*

[16] **2020**: Roughly meaning, “O three and four times blessed, those to whom it befell to encounter death before their fathers’ eyes below the lofty walls of Troy!”

*

**BOUND FOR MUD POND**

*

# Chapter V

*

Black Ducks. — A Muskrat House. — The Muskrat as a Pet. — Chamberlain Farm. — Apmoojēnegamook. — Into Pongokwä′hemook. — Evidences of Illegal Hunting. — Some Reflections on Game Protection and the Game Laws.

*

**THE** next morning, we broke camp at eight o'clock. The clouds of the day before were rolling away, and the outcoming of the sun seemed to promise fair weather. Silas took our things from camp down to the lower end of the carry, while Joe and the writer went back for the canoes and poled them down the brook. In the dead-water near Chamberlain Lake a large flock of black ducks were basking in the sunshine, and diving and playing at hide and seek. They fairly threw themselves out of the water, and back into it again with a splash, scattering it about them in clouds of spray. Their joyous "Quack, quack, quack!" added to the animation of the scene, as, unconscious of our presence, they continued their antics. The black duck is without doubt the wildest of the family *Anatidae* that comes under one's observation in Northern Maine, and it was not long before this flock, at a note of warning, a single "Quack!" from one of the leaders, fled precipitately from their playground, and that too before we were within fair rifle-shot of them. Near the end of the carry, at the edge of the brook, we saw the track of a deer, the water being still roily.

On our way through the dead-water we came upon the newly-constructed house of a muskrat (*Fiber zibethicus*),[1] or musquash, and stopped to examine it. It stood four feet from the water's edge, upon a sort of soggy turf, which sank perceptibly at each step we took. The water in the brook was much lower than the writer had seen it on many previous occasions, and lower than it ordinarily is when the Chamberlain and Telos dams are shut, which frequently happens at this season. A foot more of water would have made the house uninhabitable. The latter was four feet high and three and a half feet in diameter, and nearly cone-shaped. It consisted of bits of fibrous turf, grass, or roots, with very little mud or soil in it. Small sticks, chips, and a good deal of sedge, were mixed with the bits of turf, which were generally less than the size of one's fist. They were laid together without apparent architectural skill, and with his paddle the writer easily and rapidly shoveled off the top of the house. The interior was about eight inches high, and the floor three or four inches above the water, so that the roof under the centre of the structure might be said to be three feet thick. The apartment was crescent-shaped, and occupied rather more than one half of the space covered by the erection, being two and a half feet long by one and a half feet wide. It had two entrances, one at either horn, which led through long passageways down into the brook. The walls in their thinnest part were six inches through. The floor was strewed with sedge, and there appeared to be room enough to

---

[1] **2020**: Musquash or muskrat - are medium-sized rodents with an omnivorous diet. They are not, however, members of the genus Rattus. Also, (*Ondatra zibethicus*).

accommodate from four to six rats comfortably, although the writer has been told by trappers that they never find more than two muskrats in one house.

The flesh of the muskrat, or musquash,[2] is very tender and delicate, and makes a delicious cassambo, or stew. If the animal is properly skinned, there is no musky odor or flavor traceable in its flesh. It is often the only fresh meat the tourist can procure. He shoots it by moonlight as it swims across some quiet stream, leaving behind it a long, spreading wake, or he takes it in traps set for the purpose. The open trap is placed in a little depression made by the hand in the edge of a grassy bank near deep water. The chain is fastened securely to a stick pushed into the soil, and short twigs are stuck upright on each side of the trap to guide the rat's footsteps on to the treadle. The trap springs, catches his leg, and he jumps back into the water, where in his struggles to get free he almost always drowns.

On one occasion, while camping in Maine, the writer let his guide set three steel musquash traps, to provide our table with a change of food. The next morning, we found each trap had taken its victim. Two of the musquash were dead. The third, more of a philosopher than the others, having found escape impossible, had climbed out of the water on to a log, and had evidently passed the night there. As we approached, he jumped into the water again, and, keeping his head above the surface, resisted every attempt of the guide to secure the trap, and sprang at his hand repeatedly. Finally, by using the paddle-blade as a

---

[2] From mωskωéssω (Râle),— a name perhaps easier to swallow than the other.

guard or shield, the trap was detached, and, with the musquash, transferred to the canoe. Bunny, who was young, probably born the previous spring, we took to camp, removed the trap from his leg, and tied to the other hind leg a piece of strong but soft rope-yarn, the other end of which was made fast to a tree. Thus anchored, he slept all of that day on the ground under some boughs we had gathered for him. In the afternoon he ate, with great avidity and apparent relish, pieces of raw and boiled potato, and the writer took him down to the lake and let him swim along the shore. At first, he appeared not to use his fore legs in swimming, — a fact said to be always true of the beaver; but on a subsequent occasion he was distinctly seen to use them for that purpose. On the day of his capture and on the next morning Bunny was very docile, never seemed afraid, and allowed himself to be stroked repeatedly. In fact, he appeared to enjoy it, and soon became a great pet. But the best of friends must part. We had to leave our camp for several days, and when we returned Bunny had "cut strings" and gone. Success to his future! May he never fall into another trap!

Making our way with difficulty through the tangled mass of driftwood which choked the mouth of Mud Pond brook, we entered Chamberlain Lake and struck across towards the farm, which lay N. 22° E. from us. The lake at this point we thought a trifle short of two miles wide, but were subsequently told at the farm that it had been measured in the winter, and was two miles and forty rods.[3]

---

[3] **2020**: Forty rods - a little over another tenth of a mile.

The first clearing on Chamberlain Farm,[4] and on the lake, was made in 1846. Much of the land on and near the lake then belonged to D. Pingree, Esq., of Bangor, and the impetus given to logging operations by the successful completion of the Telos Canal rendered the establishment of a storehouse, or base of supplies, in that region, not only expedient, but necessary. The farm, which has grown to large proportions, is now owned by Hon. E. S. Coe of Bangor, and on it are raised yearly large numbers of cattle and sheep, and also potatoes, grain, and vegetables. So well do sheep thrive there, that a short time before our arrival one became so fat that, in the words of the superintendent, Mr. Nutter, they "had to kill him to save his life; couldn't lug himself around." Mr. Nutter also told us, that, when the season was not backward, he raised just as good corn as grew anywhere in the State of Maine. Good apples grow there, as we proved to our entire satisfaction.

There are on the farm, the year round, some six or eight men, a jolly and good-natured set. Woman's society is seldom or never vouchsafed them, and they are catered to by a man-cook, — at the time of our visit, a French Canadian. As this was the last human habitation we should see until we reached the lower parts of the Allagash, we improved the opportunity, and the cook brought forth from hidden storerooms a pan of rich, delicious milk, a plate of fresh cookies, and a basket of apples. The guides looked over their supply of tobacco, replenished it, refilled the potato-bag, and ground the axes.

---

[4] **2020**: Located at Apmoojēnegamook Point, commonly known as Hog Point. Formerly the operations hub for the Pingree Coe Timberlands, a supply depot established in 1846 for the lumbermen.

Apmoojēnegamook[5] is not an attractive lake. The view from it of Mount Ktaadn and the Nesowadnehunk[6] Mountains, to be sure, which grows finer as one nears the northern end of the lake, is one of which a person never tires. It is true, too, that from the high ground back of the farmhouse may also be seen the Traveller, Trout Brook, Soubungy,[7] White Cap, Baker, Lily Bay, Spencer, Lobster, Toulbah,[8] and other mountains, but the lake itself, surrounded by low, rolling, dense forestland, and its shores rigid, as it were, and covered with driftwood, is not a place where the tourist cares to stop.

We reached the dam without incident of any kind, and lunched there. At the second dam, an eighth of a mile below, there are almost always plenty of trout. The writer cast his flies over the pool a number of times, with the result of hooking and scratching the one single trout there was in it. The only things that fairly got on to his hook were two sturdy chubs.

Silas told us that during the previous year he was going through the woods near this dam, and suddenly came upon a well-known Moosehead guide, who was sitting on a log with a shotgun over his knees. The guide told Silas he was on his way down the Aroostook with some gentlemen who were camped

---

[5] **2020**: Chamberlain Lake. The lake in Hubbard's time was a logging highway and this may have impacted his harsh assessment. It is 10,932 acres with a max depth of 154 ft.

[6] "Stream among the mountains."

[7] The Indians do not recognize this word. The name they give for the pond or bog at the foot of the mountain is Alla't'wkikamōk'sis, "ground where a good deal of wild game has been destroyed."

[8] Between Caucomgomoc and Allagash Lakes. Toulbah is from Tωrebé (Râle), "turtle," and the mountains are so called because of their low, flat shape, as seen from Chamberlain Lake and elsewhere.

on the carry, and they had sent him out to see if he could find a few partridges. He'd "be dod-buttered," he said to Silas, "if he was going to chase around them woods." So, he sat quietly on his log until he thought it time to return to camp, where he doubtless reported that game was very "scurce."

The difference of level between Apmoojēnegamook and Eagle Lake is said to be twenty-one feet and six inches.[9] The connecting stream, the Allagash, is very short at this point, and its width was somewhat enlarged in 1843 by the "going out" of the original dam, built in 1841.[10] The second or lower dam is now in ruins. Its functions, those of the lower gate of a lock, ceased some years ago, at the time when Chase Dam at the foot of Churchill Lake[11] was destroyed. The level of Allagaskwigam'ook and Pongokwa'hemook was raised by the latter dam, so that logs could be taken up into the lock, and from there into Apmoojēnegamook, whence they went down the East Branch of the Penobscot.

We carried our canoes past the lower dam, and had soon run through the rapids below it, and were on Pongokwa'hemook.[12] The surface of the lake was smooth and glassy, and a haze of Indian summer hung over its distant portions, and took away that hardness which is apt to be a characteristic of many lake landscapes in the forest wilderness. The shores of

---

[9] See pamphlet, "The Evidence before the Committee on Interior Waters, on Petition of Wm. H. Smith, &c., &c. for Leave to build a Sluiceway from Lake Telos to Webster Pond," page 25.

[10] Ibid. Testimony of Shepherd Boody.

[11] Allagaskwigamook, Bark Cabin Lake.

[12] **2020**: Eagle Lake.

Pongokwa'hemook are not of that rigid and unbending nature which is peculiar to those of Apmoojēnegamook, but run out into points and projections, and form deep, far-reaching bays and coves, into which the canoeman would fain penetrate. Their borders are covered with a growth of graceful birches, which have replaced the earlier pines and spruces killed by the backflow from Chase Dam, and several islands rise out of the water in mid-lake, most of them covered with a thick forest growth. On the sloping sides of one of these, years ago, an incision was made and a budding farm ingrafted, but the wound has grown together again, and the scar is now scarcely visible.

*

**MOUTH OF NAHMAJIMSKITEGWEK**

*

As we entered the lake, on our right at its head we heard the deep baying of a hound, followed by a rifle shot. Some reckless lawbreaker was evidently at work, secure from interruption, and probably from punishment, at the hands of the officers of the law. Two miles farther on, near the shore of Pillsbury Island, we saw a quarter of moose-meat spoiling in the water.

The interest in game protection in Maine has lately received a decided impulse, which bids fair to bring about a healthy reaction. Several years ago, a general apathy prevailed on the subject, not only in that State, but in many others as well. Game laws were enacted, to be sure, but only to be laughed at, and the slaughter of game, in and out of season, went on as largely as before. Since then, however, a healthy public sentiment has been steadily growing, thanks to the efforts of a handful of persistent men and a popular journal, and humanitarians rejoice to see that the wild animals of the forest are at last deemed worthy of active and effective intercession, to prevent their speedy extermination, or their capture by methods at once heartless and cruel.

In every man, who for the first time takes up a gun to go hunting, there seems to be inborn a love of killing, no matter how refined or delicate may be his instincts in other respects, and no matter what his education and previous surroundings. This instinct leads him to take the life of wild animals, whether the dangerous wild cat or the defenseless deer, with an eagerness that is at times ferocious, and a recklessness that is appalling. He is not satisfied with the capture of one deer for his present needs, but his impulses lead him to kill others, if more be at hand, no matter if his necessities are amply supplied, and no matter if the slaughtered animals must lie in their tracks and rot. It is the same old story, on the plains and in the forests. He wanted to see if he could hit them! The pride of boasting to his neighbors that he shot such and so much game, the satisfaction with which he already sees their admiring faces turned towards him, as he pours into their straining ears the ornate and circumstantial tale of how he did it, the exclamations of wonder and envy with which they look at his trophies, — in a word, his desire to become a hero, — this indeed must be the source of that wolfish passion. We see signs of it in the public prints, haply much less now than formerly, where we are told how this man or that man, "sportsman" so called, killed thirty antelope in one week in Colorado, or how some great angler, a clergyman perhaps, caught seven hundred trout in three days' fishing. We read, too, of game-protective societies — alas! hollow name — that have an annual hunt. They divide themselves into two parties; the birds of lake and forest have each a number set upon its head, and these gentlemen amuse themselves by seeing how many of these poor creatures they can destroy, and the side that accomplishes most of this devilish work is lauded and cheered. Indeed, far from being actuated by any motives of compassion for the brute creation, these so-called game-protective societies seem to have for their chief aim and function the prevention of game-killing by others, simply in order that, at a given time, there may be more game to satisfy the greed of their own members.

This singular and unfortunate instinct to destroy game may arise in part from a desire to see and examine animals of which our only knowledge, perhaps, has been gained from books, and which have always been covered to our minds with a veil of mystery. After this desire is once satisfied, the tendency to further killing in many cases dies away, and a feeling of humanity towards the dumb creatures takes possession of us, as far at least as the gentler or innoxious animals are concerned. Then it is that our better instincts assert themselves, and unless our bread and butter depend upon it, we care little for further trophies of the chase.

The excitement which invariably comes over a person, not an experienced hunter, when he suddenly sees game before him in the woods, and the irrepressible something which makes him shoot, occasionally lead him to break the laws entirely without premeditation, and often in direct opposition to previous intent. A man, too, may be suffering for want of proper food, and may take sparingly of that provided by a beneficent Creator, at a time when his fellow-men have said he shall not do so. Certainly, there is enough charity in the world to forgive the wrong of such acts as these, if reparation be made or offered. In and by themselves, apart from their being *prohibita*, — breaches of the law, — these acts are not *immoral*; but the example they set to others, who are perhaps less particular or less conscientious, is bad. The law has been broken, and the penalty must be paid by all alike, the high and the low, the rich and the poor. Otherwise the whole fabric of game protection would be destroyed, and the law brought into contempt.

The man, however, that deliberately and with premeditation transgresses the law, as for example by using hounds or allowing them to be used by his guide, or by hunting by jack-light in the summer months, merits no charity from his fellowmen, but deserves instead their unqualified condemnation. To a much greater degree than in the case above stated is his example pernicious and demoralizing. His guide rightly says, "If this man can come here and take game when and how he pleases, why should not I have the same privilege?"

In the same category should be placed, too, the man that catches and kills twenty trout, when ten are all that he can use; or the two men together that shoot a second deer, simply for the hide or antlers, when half of one deer is all they can possibly eat.

One cause of the continued disregard of the game laws in Maine has been the want of means, on the part of the authorities, to enforce them, and another has been the state of public sentiment, which has heretofore been adverse to prosecutions and convictions. Many of the natives, in settlements near the game districts, have been so tainted with old sins of their own commission, that new breaches of the law could not be punished owing to threats of reprisal or counter prosecution. Happily, both of these difficulties have been removed, in part at least, and a new era is beginning to dawn.

Of the native population the Indians break the game laws vastly more, in proportion to their numbers, than the whites, and this is only what might be expected. It is the Indian's nature to love the woods and the excitement of the chase, and he longs for them both with a keenness which we white men, as a rule, can scarcely understand. It is said that John Pennowit, often quoted in these pages as an authority on Indian place-names, when he was eighty-six years old, and bowed with age and disease, made his preparations to go hunting in the autumn. The writer once heard a Moosehead Indian wistfully and repeatedly exclaim, while detained by head winds for several days at Chamberlain Farm, "I wish the wind would go down: I want to get into camp, and shoot a caribou or moose." This man had lost nearly all his fingers, but, notwithstanding his apparent helplessness, went hunting regularly every autumn.

Game protection consists, not in making elaborate laws, but in the persistent and courageous enforcement of simple ones. The game wardens should be, first of all, above suspicion. It is a sorry spectacle to see a warden, who acts as guide for a party of tourists, help them kill a moose out of season. Then again, the wardens must be aggressive. Their salary should be such that they can afford to devote a great part of their time, during the close season, to going over the highways of public travel through the forests of their respective districts. The woods of Northern Maine, between the east and west branches of the Penobscot, are not likely to be settled to the extent of driving out the game in them for many years to come. This wilderness is a vast game preserve, which now profits the people of Maine many thousands of dollars annually, and it can be made still more profitable, if the means are only forthcoming to protect the fish and game in it. There are many persons deeply attached to these fair lakes and streams, and noble forests, who visit them yearly and take their friends there. If they, one and all, will by their example join in the efforts to have the laws respected, and will take sparingly and humanely of the bounties of Nature's providing, it will do more towards game protection and game culture than the penal legislation of ten years.

## Chapter VI

*

The Loon considered Musically and Otherwise. — Camp on Nahmajimskitegwek. — Hornets and Maple Sugar. — Visitors. — An Excursion to Haymock Lake.— Rough Water. — Allagaskwigam'ook. — Through the Breakers. — Guides. — Indians *vs*. Whites.

*

**ONWARD** we floated over the unruffled surface of Pongokwä'hemook.[1] From the top of an immense dead pine tree on Pillsbury Island flew a bald eagle (*Haliaëtus leucocephalus*), which mounted in a graceful spiral higher and higher, until, a mere speck in the sky, it became scarcely distinguishable. Far out on the lake we heard the warning cry of a loon (*Colymbus torquatus*),[2] whose white breast, unusually conspicuous above the smooth water, seemed twice its ordinary size, — a phenomenon probably due to mirage.

The loon is one of the most interesting of the ordinary features of camp life that come under one's observation among the lakes of Maine. Every lake, and almost every pond not too small for its safety, contains one loon, at least, if not a family of them. We see them constantly, or hear their weird cries at all hours of the day, and at night too. At early morn they circle the

---

[1] **2020:** Eagle Lake – Approx. 9,500 acres. Max depth of 124 ft. It is the first, largest, and deepest lake in the Allagash River Waterway. (Source: Maine.gov)

[2] **2020:** By 1956 the genus for the loon was assigned as Gavia immer.

shores in search of food, and at these times often come close to our camp.

Or perhaps at mid-day, attracted by the unusual sight of an overturned canoe, or of a tent whose white canvas gleams through the trees, they approach to make a closer observation, for the loon is an inquisitive bird. Quietly they swim on, warily looking from side to side, now stopping in their course, or sheering off a little, then again advancing, until, satisfied, they go elsewhere about their proper business, — whatever that may be. On a large lake the loon seems indifferent to the presence of man, or even of such noisy things as steamboats, unless perhaps it is accompanied by its young. Then it is always in a flutter, and we hear its anxious "Hoo-hoo-hoo, hoo-hoo-hoo!" as if it were in great alarm. This is especially true on a small pond. The din it then makes is incessant, and resounds from shore to shore, gaining strength as it goes.

*

**2020 EDITION**

**LOON ON A MAINE POND**

Far from being a "foolish" bird, the loon often exhibits much cunning. When disturbed, the parent and its young swim off in different directions, the latter seeking seclusion near some reedy shore, while the former tries to draw attention to itself. When pursued it dives repeatedly, and on reappearing exposes only its head above water. Sometimes, too, it doubles on its pursuers, or rather under them, and often thus escapes most mysteriously from their sight. Undoubtedly the loon's reputation for being "silly" or "crazy"[3] is due to its cry of alarm, which, oft repeated by a single bird, or in chorus by several of them, may sound to some ears maniacal. This cry is certainly at times exasperating to the hunter on the lookout for game, however little in fact the latter may be disturbed by it.

Subjectively the loon seems to have been little studied, but its different notes furnish us with abundant material to interpret its varying moods. Its cry of alarm, and its shrill note when on the wing, are probably by many persons supposed to be the only ones it utters, the expression of its joy and affection being entirely overlooked. Its notes are more significant than those of many other birds, at times merry, tender, dreary, or full of fear, but almost always musical. Take the most common cry, with its several variations,

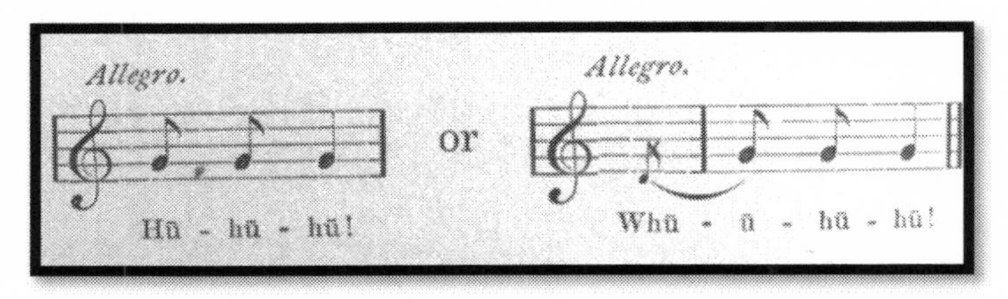


[3] Possibly a popular error, the Scotch word loun, or loon, meaning "a stupid fellow."

Each note short and crisp. This indicates ordinary surprise or alarm, while in

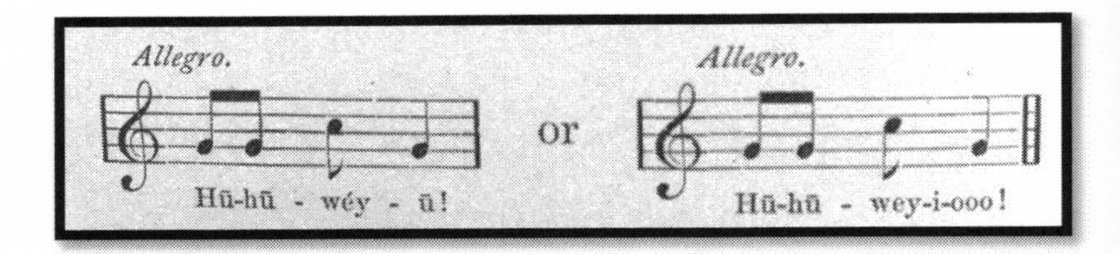


the alarm seems to be increased.

At night the camper often hears, the second and third notes slurred, and often flat,

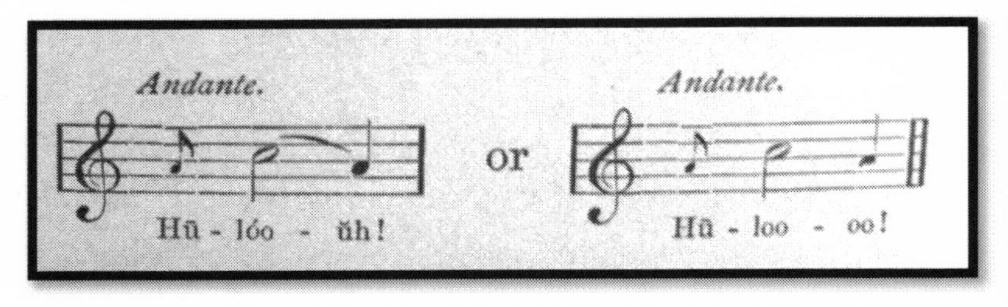


sometimes supplemented by

These dreary night cries are startling, and sound like those of a child in distress. They may be plaintive utterings to the orb of night, or the outcome perhaps of the bird's disturbed slumbers. Weird enough they seem, thus breaking the solemn stillness of the forest aisles.

Again, at night there comes softly over the water a single note, full of tenderness, like the cooing of a mother to its young, a short gentle "Hū!" or sometimes the longer "Hū-whū'-oo!" in low, plaintive tones. These sounds the writer has heard repeatedly on ponds where there were known to be families of loons, and the sounds seemed too full and mature to come from other than the old birds.

By far the most interesting manifestation of the loon's feelings, however, is the cry, sometimes heard late in the summer, but oftener in the springtime, in that joyous season when all nature is bright, and our bird, happy in his old haunts perhaps, or with his mate or new-born offspring, rings forth with a merry swing his

*

*

Who will say, then, that the loon has no feelings kindred with our own, consign him to a place among the insensates, and let his only function be to serve as a target for the wandering bullets of summer tourists?[4,5,6]

After having inspected some camp grounds on the island, and not having found them to our liking, on Joe's recommendation we went to the mouth of Smith Brook,[7] where we found a very good site. Unfortunately, it appeared to be occupied, or partly so, for a tent stood on it, in which were a few cooking utensils and some clothing, and appearances generally indicated that the owners had been gone for a day or two, and might be expected to return at any moment.

Among other things, in plain sight from the lake, were three wooden frames made by Indians from the stems of young birch-trees, and on which evidently had been stretched and dried the skins of some large animals. To a tree opposite the tent, and about four feet from the ground, was nailed a small box, the open end of which faced sidewise, instead of upwards, — a covered shelf. In it was a quantity of maple sugar. An attempt to take out a cake of the sugar disturbed and dislodged a horde of hornets, which, secreted on the inside, had been feasting on the sweets. Truly, "there is no new thing under the sun." Sugar and hornets in combination, in the woods as well as in the outer

---

[4] It must not be supposed that the musical expression of the loon's cries here attempted, is exact and accurate. The writer simply wishes to give a general idea of those cries, as nearly as he can reproduce and represent them.

[5] **2020**: What might Hubbard think today when one can find videos and sound files of the loon's songs with a simple online search.

[6] **2020**: In the 1800s loons were regularly hunted. This became illegal with the 1918 federal Migratory Bird Treaty Act. Now, the only thing tourist (and residents alike) shoot with, thankfully, are their cameras.

[7] Nahmajimskitegwek, "the dead-water extends up into the high land."

world. Only the sugar is visible until it is stirred up, when the hornets come to the surface. There is one difference, however. Here in the wilds the two components, sugar and hornets, may be separated, and the latter driven away.

We fancied we knew the owner of the strange tent, and as he was a friend, we forthwith pitched our tents alongside. A light rain had begun to fall, and a canopy was improvised over the rough table, at which we were soon eating supper. In the midst of our meal, the clatter of a paddle was heard on the rocks at the landing, and in a moment the face of a stranger peered at us through the dimness of our candle-light. Thinking him a belated canoeman in search of shelter, we bade him welcome, told him to make himself perfectly at home, and invited him to a share of our supper. Imagine our petrifaction at learning that *he* was the owner of the strange tent, and of course by courtesy the lawful possessor of these premises, *pro hac vice,*[8] as the lawyers say, while we were really trespassers. That made little difference, however, and we were soon on very good terms, — a matter of course among gentlemen that fall together in the woods.

The next day was devoted to an excursion up Nahmajimskitegwek to Haymock Lake.[9,10] Sartor preferred to remain behind, and said he could get his dinner himself; so, both guides accompanied the writer. The day was warm and bright, and Nature looked her best. When once over the various

---

[8] **2020**: Lawyer term for, “for this occasion.”

[9] This name was probably taken by white men from the last two syllables of the Indian name for Eagle Lake, "Pongokwä'hemook."

[10] **2020**: Haymock Lake, 900+ acres, max depth of 61 ft.

beaver dams and above the carry, we found a long stretch of dead-water. The stream widened, and its sluggish course took it through a hackmatack bog. The bright green of the larches, fading away in the clear sunlight to a yellowish autumn hue, was in strong contrast with the darker green of the spruces, while the scrub-willows and other bushes that covered the bog, had put on a reddish-brown, that be tokened the approach of winter. On the way up, Joe pointed out to us a spot where he had once "called " a moose for a gentleman, but the old bull, after having come almost within sight, grew suspicious and went off.

On our way back to camp we picked a quantity of swamp-cranberries, which grew in abundance near the mouth of the brook. That evening the Captain gave us a funny account of his attempts to cook dinner, and showed us as the result of his day's work three very pretty little sketches. We afterwards made observations, by which we calculated roughly that the magnetic variation of the compass for this place was twenty-two degrees.

The next morning, we left camp in the face of a rising wind. Rounding the point below the mouth of the brook, we soon came to one of the widest parts of the lake, across which the wind had full sweep, coming broadside on. The waves grew momentarily more threatening, a white-cap every now and then dashing against our bows, and breaking in to our laps. While we had no fears for our personal safety, there was great apprehension lest the flour and other provisions might become wet and unfit for use. By taking the waves on the quarter, however, and paddling with all the energy we possessed, we managed to reach a quiet haven behind a bluff at the narrows, and, stepping out upon a beautiful little sand-beach, we soon dried our wet garments and bailed out the canoes. At this point, which is in many respects the most attractive one on Pongokwä'hemook, the water was as serene and placid as if no wind were blowing; and ere we were ready to go on, we had persuaded ourselves that the wind we had lately felt was only a squall, and that our further passage would be smooth.

*

**AMONG THE WHITE-CAPS**

*

In this, however, we had mistaken; for on leaving the narrows and entering mid-lake again, we found the wind quite as strong as it had been, and the sea even rougher than before. Our course this time, however, lay more with the wind, and, by keeping the canoes diagonally across the trough of the waves, we were carried speedily along with the latter, and soon reached the thoroughfare. Here, on the westside, about a mile from Thoroughfare Brook,[11] and sheltered from the wind, we stopped for lunch.

This thoroughfare, about two and a half miles long, connects Eagle Lake with Churchill Lake, and, with the exception of a slight current at the outlet of the former, is virtually "dead-water." On its west side the land is mostly low and flat, and was overflowed to a great extent at the time Chase Dam was in use. As a consequence, a forest of dead trees, or "rampikes," bristles on that side, and makes a picture of dreariness. But this was moose-ground, and as we continued on our way past it not a word was said, and all eyes intently scanned the grassy undergrowth on our left, in the hope that we might see some game. We were now in October, and our appetites were whetted not a little by this fact, but perhaps more by the fact of our long and enforced abstinence from fresh meat. The Captain's formidable revolver lay before him, ready to be called in to use at any moment. This time, however, no moose exposed themselves, or if they did, we did not see them.

We passed the mouth of Thoroughfare Brook, and in a few moments more were at the head of Allagaskwigam'ook, Allagaskwigam'ook or Churchill Lake. The wind during the past few hours had increased greatly in intensity. Our course now would lie directly with it for nearly three miles, and although the water immediately about us was pretty smooth, we knew there must be a heavy surf on the beach at the lower end

---

[11] Sahbimskitegwek, "a branch or stream that empties between two large bodies of water."

of the bay, where we wanted to land; and it became a serious question, whether, even if we should escape swamping before reaching shore, we could even then land without injury to the canoes, or without being over whelmed by the breakers. Silas, who never was perfectly at ease on an open lake on a windy day, said he thought we would better not go on, but rather camp near where we were. Joe was non-committal, but finally said we could make the attempt; and as the prospect of losing half a day was not over pleasing to the rest of the party, the word was given, and on we went.

For more than two miles, and until we were within a quarter of a mile of the shore, all was delightful. The wind drove us onward without much paddling on our part, and we were rapidly closing up the gap between ourselves and the land; but now the waves became formidable indeed, and the cove before us seemed a mass of mad, seething breakers. Retreat was impossible, and we looked each at the other's canoe, to see how it was bearing the strain, and shouted words of mutual encouragement. As each on-coming wave seemed about to engulf our canoe, Silas swung the bow around, and we were lifted up and forward safely, with the shipment of only a little water. Joe's canoe was not so fortunate, for the "gunwales" were lower, owing either to the form of the canoe, or to its being loaded more heavily than ours, and two or three wave-tops swashed over the side and threatened to swamp it. We were now almost on the shore, and laying aside his paddle, with one hand forward on the rail, the writer crouched in his place ready for a long spring. The moment arrived at last, and as he felt the canoe touch the sand, he jumped forward, grasped the bow and pulled with all his might, while Silas, taking advantage of another wave, gave one more push, and in a moment more we had the craft high and dry on the sand, and were gratified, on looking around, to see that the others had made an equally successful landing.

It was now half-past two o'clock, and although we ourselves had hoped to go farther, the guides evidently had no idea of doing so, for they proceeded to pitch the tents and get firewood for the night. Each of them subsequently took a load over to Marsh Pond, and made some betterments on the carry, so that no time was really wasted. In this particular instance neither the Captain nor the writer made any remonstrance nor asked any questions, as we preferred to see how things would turn out, and waited to assert our authority on some occasion when interference might be more necessary.

The experience of having two guides in the party was somewhat new to the writer, as for several years previously he had made his forest excursions alone with his faithful Silas. However much a man may be inclined to rely on his guides, and follow their suggestions, he does like to have his wishes consulted by them. With two guides, especially if they speak a strange language, a man feels very much like a figure-head. Even if conscientious in other respects, they are apt to lay out their own programme, and by well-timed suggestions, or, if these fail, then by what is known as "hanging back," unconsciously force one to carry it out. The writer does not mean to have it inferred that his guides on this trip resorted to either of these bad practices, but make plans from time to time they certainly did, — a thing in itself quite natural, and not necessarily objectionable.

Hanging back, in its worst sense, is never resorted to by competent and faithful guides. They are up before sunrise, attend to the daily repairing of their canoes early in the morning, during enforced pauses between other duties, and are quick about packing up and leaving camp. They paddle briskly, and keep it up until a reasonable and seasonable hour, and in turn expect reasonable treatment from their employers. But the hang-back guide, on the other hand, gets up in the morning when he happens to feel like doing so. He builds the fire leisurely, and wonders what to cook for breakfast. He overhauls the entire kit, and, by the time the meal is ready, eight o'clock has come. After breakfast he finds that he has no hot water for dish-washing, and, while the pot is waiting to boil, he lies back leisurely smoking his pipe. Meantime the tourist has perhaps done much towards getting the luggage packed, and, when all is finally down at the landing, the guide suddenly remembers that his canoe needs "pitching." The embers of the camp-fire must be brought to life again, the pitch melted, and the canoe-bark dried. These delays and this dawdling are vexatious in the extreme, and in some few cases are indulged in for the sole purpose of prolonging the period of the guide's engagement, or of preventing him from getting too far away from home and into the neighborhood of carries. He will try to persuade you by loud talking, embellished with oaths, that certain places are too difficult of access to be reached; and if you insist upon trying, the chances are that he will lose his temper and become sulky. Again, some indispensable article, as for instance the axe, will be left behind, and its absence accidentally (?) discovered two hours later, when the party is not too far advanced to send back or to go back for it.

Pages might be written on the many deceptions practised by men who profess to be honorable guides;[12] — of engagements broken on receipt of better offers from elsewhere; of demands for higher wages than those previously agreed upon, — at a time, too, when to refuse the demands would be disastrous to one's plans; of deliberate falsehoods told to avoid a little hard work; and of a thousand other ways of embittering the visitor's stay. Heaven help the man that falls into the clutches of such guides as these! Happily, there are but few of them left, in the Moosehead region at least.

Another species of guide often met with is the honest, good-natured, but improvident fellow, who always fills the teapot up to the brim for every meal, and with a decoction[13] strong enough to shake the steadiest nerves, although he knows that not more than two thirds of it will be drunk,[14] and then

---

[12] **2020**: Licensing for Registered Maine Guides began in 1897. In the first year 1700 individuals were licensed. For decades there was no standardized testing for becoming a Maine Guide. "Licensing" was under the authority of a game warden to determine if the applicant was considered qualified. The first standardized testing and certification procedures to be a Maine Guide began in 1975. As of 2020, there were approx. 6,300 licensed guides in the state with various specialty classifications. (Source: Maine.gov) The first Maine Guide was Cornilia Thurza Crosby, or "Fly Rod," as she was popularly known.

[13] **2020**: Resulting from concentrating the essence of a substance by heating or boiling.

[14] The writer says this advisedly, although he is aware that it takes a pretty big pot to exhaust the tea-drinking capacities of the average back woodsman.

deliberately, but invariably, throws the remainder into the fire. The camp table before each meal groans beneath its generous load: after mealtime it often looks as if a famine had swept the land. You and your guide eat abundantly and heartily, after which the guide dumps on his own plate the food that is left, and eats it, in order that it may not be "wasted!" or, more rarely, he puts it carefully into a corner somewhere, and on the next day throws it away. If you remonstrate with him, he will smile pleasantly, and repeat his improvidence at the earliest opportunity. Your provisions vanish long before the proper time, and, if your eyes have not witnessed the extravagance that was going on daily, you mentally blame the person who told you what quantity of food to take with you. The result is that you reach your goal, and, instead of having a quiet rest there, you must hasten back to avoid starvation.

The writer's experience in Maine forests, both with white and Indian guides, extends over a period of thirteen years. As canoemen on rivers, the Indians, as a class, cannot be surpassed. As hard workers and willing workers, the few who have served him have been without equals; but it is a fact worthy of notice that they all had some white blood in their veins. As a rule, the Indian lacks volition, the absence of which is often as objectionable as the presence of its opposite, too much self-will. Putting the two races side by side, there need be no hesitation in saying that an experienced Indian hunter, strong, willing, and respectful, is the most profitable and instructive, and withal the most agreeable guide that one can have. Sincerity bids the writer add, however, that these characteristics are not commonly found united in the same person, white or Indian.

Into Allagaskwigam'ook, within a few rods of our camp, flowed two brooks, one on each side. They are called the "Twins," and the more northern one comes from Spider Lake, or Allagaskwigam'ooksis.[15] This brook is usually too low to make canoe-navigation profitable, and parties going up it get along faster by using the carry which runs along its left bank. On the afternoon of our arrival at Allagaskwigam'ook, Sartor and the writer strolled over the carry. There were few signs of recent travel there, and we did an hour's work "bushing" out the path, in order that our labors might be less severe on the following day. The guides, after looking the ground over, decided that they would try to get up the stream with the canoes, and would first carry the "stuff " over on their backs.

[15] The diminutive of Allagaskwigam'ook.

## Chapter VII

*

The Indian Pack. — How to get over a Carry. — An Artist in the Air. — Through to Allagaskwigam'ooksis. — In Camp. — Acquaintance with the Chubs. — Lost in the Woods. — Mahklicongomoc. — Carrying a Canoe. — Canvas *vs*. Birch.

*

**WE** awoke early the next morning to find a heavy mist hanging over Allagaskwigam'ook.[1] The sun soon broke through it here and there, making fantastic forms and fairy-like imagery. Now, part of the denser cloud-mass would momentarily roll aside, and indistinctly expose to view the tops of some distant pine trees, which grew as it were out of the vapor sea, and whose mist-bespangled branches gave back a thousand glittering rays. Again, a mist-bow appeared, in the form of the rainbow, but without its colors; and finally, the whole mantle of fog rose and floated away before the morning breeze.

After breakfast Sartor went over the carry alone, leaving the rest of us to follow later. As this was our first carry of any consequence where we should have no horses to help us, the camp equipage was more carefully arranged than before, and the guides made up two packs, the favorite means among the Indians of transporting their effects through the woods.

---

[1] **2020**: Churchill Lake – Approx. 3720 acres. Max depth of 62 ft. Second lake in the Allagash River Waterway. (Source: Maine.gov)

*

**JOE MAKING A PACK.**

*

An Indian pack is a heterogeneous thing, and, like the historic mince-pie, it is generally composed of a little of everything there is on the premises. The tent usually forms the groundwork, and across it on either side of its centre are laid the two ends of a long double strap, — which may be made of leather, or of a piece of the inner bark of the cedar, — tapering from the centre to each end. These strap-ends are laid a little farther apart than the intended width of the pack, and in parallel lines, leaving a margin of tent more than a foot wide outside of each of them. The margins are then folded over the straps, and may or may not meet or overlap along the centre of the tent. We now have spread out before us what for purposes of this description may be called the "pack-cloth." It is long and narrow, and at its upper end we see a wide, continuous strap, which extends from side to side, and disappears at the corners under its folds. The strap then runs along the sides of the cloth, concealed from view until its tapering ends come out at the two lower corners. On the middle of the pack-cloth are now piled buckets, blankets, pots, pans, shoes, socks, and anything else that has no more appropriate place, until a load is accumulated larger than the body of the carrier, and of a weight sufficient to tax the strength of two ordinary men. These different things are all arranged so that no uncomfortable projections shall chafe the carrier's back.

The next step is to fold the two ends of the pack-cloth over the articles just piled up, so that the structure may have somewhat the shape of a barrel with head and bottom knocked out. The Indian now usually stands astride of his pack, holds firmly with one hand the central part of his strap where it disappears among the folds of the tent, and pulls hard upon its corresponding end, which by the previous act of folding has been brought up and opposite to it. What was the side of the pack-cloth now becomes the end of the pack, and under the pulling process soon looks like the mouth of a bag, which is made fast by a knot in the strap. The other side is treated in the same manner, and we now have a shapely pack, with ends tightly closed. Along the top, from end to end, runs the broad part of the strap, and from the knots at the extremities of this broad part run the two long tapering ends. These are brought together under the centerpiece, crossed, and carried around the middle of the pack, where on the opposite side they are tightly knotted. The pack now is a firm solid mass, and the Indian, often unable from its great weight to lift it alone upon his back, either drags it to some log or mound, or by the aid of another person succeeds in getting under it. The broad part of the strap passes over his forehead, and sometimes, as an additional aid, a second strap passes from the pack around his chest.

Not satisfied with a single pack, the Indian will throw a bag or box on top of it, and trudge along over a carry without thinking of resting until its end is reached. Accustomed to these burdens on their hunting excursions, from childhood up, no wonder many of the Indians have immense necks and shoulders. Silas's development in this particular, as before stated, was particularly noticeable, and often called forth admiration.

It is said that the disposition or individuality of a person is exhibited in its true colors much sooner on a journey than on any other occasion. If there is any one kind of journey that is more trying to the temper than another, and more apt to call out this exhibition, it is a journey over a disused carry on a hot day, when one is weighted down by a promiscuous load of camp equipage. The subject is already too trite to be described anew in these pages. There are two ways, however, of making a portage, — a proper and an improper way. The latter is to take a heavy load of differently shaped articles, and to try to carry it over logs and slippery roots, or through a tangled undergrowth, either without stopping at all, or with but few rests. The strain thus put upon the body, especially of a person who is little accustomed to hard physical labor, is so great that exhaustion is likely to follow. Much the more proper and rational way is to divide one's allotted portion of the luggage into several reasonable loads, each one of which should consist of articles of the same general nature, that may be carried together without interference. Take the first of these loads, and carry it say two hundred yards. Then come back for the second, and carry it a hundred yards beyond where the first was left. If there be a third, carry it in turn a hundred yards beyond the second. Then get the first and carry it a hundred yards beyond the third, and soon to the end of the carry. It will be found that during each return of two hundred yards the body cools off somewhat, and the muscles are relaxed and rested. Moreover, by the system of altering the character of one's load, an additional rest is given to the muscles; for no one set of them is called into play constantly, except perhaps those of the lower part of the body, which can in most cases better bear the strain of an extra amount of walking than can the others that of an added weight.

Fortunately for us at Allagaskwigam'ook, the day was bright, but not very warm, and the carry, which was dry and much improved by our work of the afternoon before, gave us little cause for complaint. On the writer's arrival at Marsh Pond a strange sight met his eyes. Sartor, his legs cased in rubber boots, was far out in the pond in an attitude of sitting, but on what no mortal man could see. Before him in his hand he held a block of cardboard, while every minute or two his brush would dip into the pond as naturally as if he were painting the universe in mid-air, and this pond were his water-pot.[2]

---

[2] On the 17th of October, 1879, an autumn noted for its warmth, the writer visited Spider Lake. The surface of Marsh Pond was covered with the white flowers of the waterlily, and the air filled with their perfume. On his way back, four days later, the pond had in many places a sheet of thin ice over it, and not a lily-blossom was to be seen.

When the guides came with the canoes and the Captain had descended from his aerial perch, we embarked for the other end of the pond, which was little more than three eighths of a mile distant. We were soon separated, our respective guides fancying they saw the better passage through the shallow water. Each canoe-load looked very picturesque to the occupants of the other, as it glided along through the mass of bristling yellow-green, our heads above the reeds and sedge, while our bodies and the canoes were almost entirely out of sight.

Entering the brook once more, we passed between low grassy banks, beneath overhanging cedars and soft-clad graceful hackmatacks where they arched the stream. But the water soon grew shallow, and Sartor and the writer had to take to the woods on the right, and follow a blind trail for a mile up to Allagaskwigam'ooksis. Here at a dilapidated and picturesque old dam we awaited the arrival of the guides, who, when they came up, told us that shortly after we had left them they heard several caribou in the woods on their left, but made no attempt to "call" them back, as the rifle had been taken by us on our walk. Joe said he could call caribou, and, putting his hand in front of his mouth, he made a gruff sort of grunt, to show us how it sounded.

Allagaskwigam'ooksis is a pretty lake, in the labyrinth of whose coves one is seldom sure of being in the right place. Its waters are dark, and give one the impression of being very deep, an impression that should seem to be confirmed by the presence in them of numbers of large togue. From its surface several mountains on the north and northeast are visible.

*

**A HEAVY LOAD**

*

In deciding where to camp, the question came up as to where the old carry to Pleasant Lake[3] left Allagaskwigam′ooksis. Silas and the writer on their last visit here had found a path over the ridge which lay between the two bodies of water; but Joe said that another and a better one left the upper end of this lake, and

[3] Mahklicongo′moc, "hard-wood-land lake."

accordingly to that point we proceeded, and camped among some cedar trees in a small, deep cove behind an island. In the afternoon a light shower of rain fell. A great deal of interest attached to this camp, for we were now on the border-land between the old and the new, the tourist-trodden and the tourist-untrodden, and the pleasure to be derived from peering into strange nooks and corners in this wild country, and the likelihood of finding some enchanting lake, or of penetrating into the heart of some remote beaver colony, or into some lonely bog where the moose roamed undisturbed, quickened our expectations to a high degree.

That evening Sartor and the writer paddled around to the mouth of the principal brook that comes into the lake, and up its dismal course for some yards. A musquash startled us by flopping over in the water, with a sudden splash, near the bow of the canoe, and once or twice out of the solemn hush of the forest we fancied we could hear the movement of some large animal. The moon shone brightly on the placid water, its lustre reflected from the lily-pads on our right, while on our left black obscurity held undisputed sway.

On Allagaskwigam'ooksis we camped two days. The old carry, in many places no better than a "spotted line" used in the winter by hunters, when four feet of snow cover the worst obstructions to summer travel, had to be cut anew in many places, over a distance of a mile and a half. We were glad enough to stay in one place more than one night, as the daily routine of packing up and moving had become somewhat monotonous.

There was little in our camp life here of special interest. We paddled around the lake among the intricacies of its broken northern shore, and visited a small pond on the south. At our landing place a lot of chubs soon congregated, attracted doubtless by the bits of food which found their way into the water there. These fish grew intimate with us, and, if a hand happened to be dipped into the lake, they would swarm around, waiting for the expected meal. With them the writer has had many hours of amusement during his camping experience. They have at times grown so tame and fearless, even after but a day's intercourse, that they would crowd around his hand, lowered palm upwards among them, and snatch and pull at the soaking bread-crust which he held between thumb and forefinger. At these times he could lift them by the handful out of the water, when of course they would at once flop back again, their fall scattering for a moment in affright their legion of comrades. Reassurance would come again, though, and the scattered hosts would soon be as busy as ever in their eager and jostling search for food.

One morning while Sartor was busy in the tent and the guides were at work on the carry, the writer determined to go over to Mahklicongomoc. Thinking to find the "spotted line" quite traceable, he took no compass with him, and after leaving the guides followed a north east course. For some time, the trail was good, but it soon became blind, and the writer was compelled to depend upon his own knowledge of wood-craft, which, up to that moment he had considered very fair. He had not gone more than half a mile in this way, when to his surprise and pleasure the lake lay before him. Approaching the shore and peering through the trees, he recognized at once an island he had seen on his last visit there, in 1879. After a short rest, he turned to walk along the shore, and suddenly came upon fresh human footprints. While wondering how any person could have come there without his knowing it, he was still further startled to see beyond him a tent. Then he began to appreciate, in a slight degree perhaps, the sensations of Robinson Crusoe. "Campers here, eh?" he ejaculated, and, walking towards the tent, what should he see before him but — Captain Sartor! Well, the writer had not a word to say. He was dumfounded. He had often heard of persons, lost in the woods, wandering in a circle, but here was a case of a person, not lost, doing the same thing unconsciously, — a veritable paradox. The writer's companions of that trip will learn of this adventure for the first time through these pages. Mortification and thankfulness at his escape have made him keep the secret from them. Moral, never go into the woods without a compass, — a good compass.

The qualification in the last words is suggested by the following amusing occurrence, related to the writer by one of the participants in it. Three men were hunting together in the Maine forests, and after a time decided to return to their camp. They knew in which direction by the compass to go, in order to reach it, and by a strange coincidence they all thought north lay facing them. Each had a compass in his pocket, but one of them said that his was out of order, and he knew it did not point correctly. The needles, however, on being compared, all pointed alike, and, oddly enough, according to the preconceived notions of their owners placed north just ninety degrees out of the way. The three hunters were not long in coming to the conclusion that *all of the compasses were wrong*, and they followed a contrary direction only to become more hopelessly astray. This incident serves to confirm the wisdom of the oft-repeated advice, that, when a man is lost in the woods, the best thing he can do is to sit down, calm himself, and then follow the dictates of reason, not of blind prejudice.

Many persons, doubtless, have had a similar experience to that of the writer, when paddling over a lake in a dense fog. If the canoe takes a straight course, as when the stern-man steers by the compass, to the bowman it seems as if they were swinging around in a wide curve.

On the morning of our second day at Allagaskwigamooksis, the guides took one load of luggage over the carry, on the other end of which they still had some cutting to do, and by prearrangement were to return and eat dinner before we should move our camp. Sartor and the writer spent the intervening time in paddling around the lake, and on our return found the following note from Joe, to wit: "We are going to take the dinner at the other end of the carry." Accordingly, we shouldered our loads, and in three quarters of an hour were on the shores of Mahklicongomoc.

After dinner, while the guides went back for the last loads, Sartor and the writer walked along the shore, around to a picturesque point that ran out into the lake near some pretty islands. The mossy ground under us was literally cut up with the tracks of caribou and deer, but the only living creature that came before our eyes was a solitary loon far out on the lake. We stretched ourselves on a mossy bank under a graceful birch tree, and drank in the delicious air and sunshine to our heart's content.

The carry from Allagaskwigamooksis comes out in a broad bay well towards the eastern end of Mahklicongomoc. After a good rest, and leaving the Captain to his own meditations, the writer wandered back to meet the guides, Joe with his pack and Silas under the canoe. The canoe looked very oddly in the distance, as without visible motive power it cleft the over-arching bushes, which scraped with a harsh grating sound against its sides. Then appeared the body and booted legs of the carrier, who was unconscious of any one's approach until we were almost abreast.

When the Indian is about to take his canoe over a portage, he ties the paddles firmly along the upper side of the three middle thwarts, parallel with the length of the canoe, and a little more than a foot apart, so that they may rest on his shoulders when he is under the inverted boat, and take some of the weight from the back of his neck. The neck otherwise supports the entire burden, the middle thwart pressing against it sometimes with a sharp and cutting edge. This discomfort is partly removed by a cushion improvised from a coat or other garment, or by tying to the thwart a thin, flat, and broad piece of cedar, which rests directly across the neck and shoulder-blades, and distributes the weight over a larger surface. A strong backwoods man will pick up his canoe by the rails in front of the middle thwart, and swing it over his head with ease, while the average tourist, if he ever attempts to carry one, prefers to rest the bow against the low branch of some tree, and then get under the canoe at his leisure.

*

**CARRYING A CANOE**

*

The ordinary birch-bark canoe of Maine today is about eighteen feet long, three feet wide in the middle, and a foot deep, and the ends curve more or less according to the fancy of the builder, but seldom according to the fancy of many illustrators. It weighs from eighty to one hundred and ten pounds, sometimes exceeding the latter figure, especially after having become water-soaked. Canoes are made for hunting which have the ribs an inch or two apart, and weigh sixty or seventy pounds.

There are two principal models of canoe, the flat-bottomed or lake-canoe, and the round-bottomed or river-canoe. It goes without saying that the former is the steadier and safer of the two, and the writer considers it the better even for rapid water. The round-bottomed birch, to be sure, possesses the quality of responding at once to the stroke of the paddle or setting-pole, an advantage often of great moment in rapid water, and not to be lightly esteemed. On the other hand, this form of canoe draws more water than the other, and consequently is more apt to be scraped and injured by hidden rocks, and there are seldom passages between rocks so narrow that either canoe cannot ride through with equal ease. A flat-bottomed canoe, if made to curve somewhat on the bottom towards each end, will possess the advantages of both models, with few of their disadvantages. It will be safe, and easily guided.

The canvas canoe which the writer used on this voyage was flat-bottomed, and built after the ordinary model of the birch, except that the keel began to curve too short a distance from the ends. When loaded, the canoe had a great length of keel under water, and in rapids the current would act with such force on either end of it — on the stern going down and the bow going up— that its proper guidance became a matter of very great difficulty, and often taxed severely the poler's strength. In other respects, this canoe was admirable. Unlike the birch, it needed no repitching, and though it received some pretty hard bumps and scratches, it did not leak a drop during the entire voyage.

We camped that night at the end of a cove at the northwest extremity of Mahklicongo′moc,[4] a cove in which the water was very shallow and the bottom muddy, and from which marsh-gas rose in abundance. Fortunately, for us the wind carried the offensive odor away from our camp, and we found quite near us a cool and sparkling brook, from which we drew our water supply. This was the nearest point on the lake to Harrow Lake,[5] and an old and bush-grown logging road ran to the latter through flat and swampy ground. This road the guides set about clearing out the next morning, while Sartor and the writer went out in a canoe on a tour of exploration.

---

[4] **2020**: Mahklicongo′moc – Pleasant Lake.

[5] Megkwakangamocsis, "marsh pond." **2020:** Harrow Lake – Approx. 403 acres. Max depth of 30 ft. (Source: Maine.gov)

## Chapter VIII

*

Pine Forests. — A Caribou. — Another Disappointment. — Hornets *vs*. Flies. — The Beaver. — His General Appearance. — Disturbing a "Bachelor." — The Beaver's Habits. — His Dams. — Mechanical Skill. — Wariness.

*

**SINCE** our arrival on Allagash waters we had been struck by the great number of pine trees which lifted their heads over the surrounding forests; and this was all the more noticeable because of the general absence of those trees theretofore along other parts of our route. Indeed, the pine has almost disappeared from Northern Maine, having long since succumbed to the woodman's axe, except in this and some other sections of the State which are tributary to the Allagash and St. John Rivers, and even here the trees are mostly what are called "saplings." It is now many years since the logger sought out in this region the giant trunks of the pine, whose forms he laid prostrate, and, cutting off their buts, left the rest to waste away. The writer has often seen these immense remains, which would measure nearly three feet in diameter, stripped of their mouldering bark, their trunks almost as sound as when they fell, forty years ago. Old and decayed dams, picturesque in their ruin, totter now where years ago they guarded their resistless hoards of water, once ready at the builder's nod to let seething torrents loose down the beds of quiet streamlets, and turn their laughing ripples to a roaring flood. Driftwood of wandering logs and mighty trees lie piled and wedged about those aged gates, making desolation more desolate, while here and there stand forests of the past, lifting their naked spires heavenward, monuments of the power of water to destroy.

The forests of Maine no longer yield up their bounties with such a lavish hand as formerly. The encroachments of the logger have robbed them of their larger trees in regions near the settlements, and now he needs must seek elsewhere. The time is not far distant when the tourist will see Chase Dam resurrected, and the forests now haunted by the moose will ring again with the chopper's axe, and wilds seldom trodden save by the trapper and hunter will be strewn with the tangled *debris* of spruce tops.

Upon the day in question we skirted around Mahklicongomoc, pried into every hidden nook which might contain some novelty, and searched diligently for in-flowing brooks. The only one we could find greater than a rivulet emptied into the northeast end of the lake; and as we approached the little cove before it, we saw, standing in the shallow water some two hundred yards away, a caribou. The wind was unfavorable, but we hoped the air-currents might rise before they reached our game, and we quickly steered behind a rock, up whose steep sides the writer succeeded in climbing with his rifle. At last we were to have fresh meat, thought he, to replenish our rapidly failing stores. How good it would taste, to be sure! and at the thought our mouths fairly puckered, as did the Widow Bolte's, in "Max and Moritz,"[1] when she went down stairs for some sauerkraut. And Sartor wanted the hide, to lay beside his bed. How warm it would feel to his feet, next winter, when he should rise on cold and frosty mornings! Alas! vain hope. The wind proved recreant, and the writer had scarcely reached his perch, when the familiar signs seen on Lobster Stream were repeated, the head moving from side to side near the surface of the water, and in a twinkling the white part of the tail flopping a signal of farewell, as the animal jumped into the woods and disappeared. Regret was useless, but, swallow our disappointment as philosophically as we might, there was nevertheless traceable in the voice of one member of that party a certain huskiness, which indicated a difficulty in the swallowing process.

The next morning opened with the wind in the south, and light showers of rain falling at short intervals. The guides had finished their work on the carry, and we had hoped to leave early for Megkwakagamocsis, but decided to await the outcome of the weather. The early hours after breakfast were spent in various ways, the writer during part of the time watching the warfare waged by a large hornet (*Vespa crabro*) against the swarm of house-flies that filled our tent. These domestic flies, if indeed they be the same that haunt our firesides at home, introduce themselves to the camper almost as soon as his tent

---

[1] **2020**: Max und Moritz are two naughty little boys in an 1865 story by Wilhelm Busch. The Widow Bolte is one of the victims of the boy's jokes. Busch was a German painter and poet who wrote satirical picture stories. Given Hubbard's study of the German language, it is no wonder he would reference this story.

is fairly pitched, and the news of his arrival is quickly heralded among their relatives, near and distant, in both senses of those terms. A visitor to the tent almost as frequent as this fly, but not so "numerous," is the hornet, a "character," that, however much he may resent the intrusion of others into the neighborhood of *his* abode, in *theirs* demeans himself with decorum, and confines his efforts to attempts at making life miserable for the flies. He booms along and into their midst, and now and then, by great good luck, catches one, which he proceeds forthwith to dismember. Hanging quietly, head downwards, from the ridgepole of the tent, he turns his booty round and round, and we see first a leg fall off and then a wing, until, reduced to the semblance of a little ball, what remains of the fly is carried in triumph to the family nest. As a marksman the hornet cannot lay claim to great distinction. He seems to buzz around at random, often pouncing incontinently upon a darkened spot of mildew, and again hovering over a fly not two inches away, without appearing to notice him. Nor does his presence seem much to trouble the peace of mind of the flies, for they not only do not scatter at his approach, but even seem to jostle him. The old fellow, however, is persistent, and from among the multitude of his prey, by patient efforts, generally succeeds in taking off his victim.

Of all the fur-bearing animals that the tourist still finds in Maine, perhaps none excites so much interest as the beaver (*Castor canadensis*). Endowed with what philosophers have termed a wonderful instinct, he has long been an object of admiration, if not of reverence, among the Indians, whose oft-repeated tales of his saga city, still further exaggerated in the narratives of early travellers in the New World, invested him, until the end of the last century, with most marvellous powers, mental as well as mechanical.[2] Even now there is a wide-spread misapprehension of many of his habits, and a tendency, it may be a just one, to rank him in intelligence above many others of the vertebrates, with whose habits we have, and from the circumstances of the case can have, much less acquaintance than with his.

The beaver of Maine today grows from two to three feet long, and is covered with a thick coat of fur, the part nearest the skin being of a light brown color and very soft, while the longer hair varies from a chestnut-brown to a dull, dark slate, and completely hides the shorter hair from view. For domestic use the skins are generally plucked of the longer hair.

The weight of a three-year-old beaver varies from thirty to fifty pounds. The body is thick and full, the head, eyes, and ears are small, the nose oblique, the fore legs short and their feet digitigrade, the hind legs longer and the feet webbed and plantigrade. The four front teeth or incisors are two or three inches long from base to tip, and are composed of a very hard dark-brown flinty outer substance and a softer inner material which wears away more rapidly than the former, and leaves the teeth with sharp bevelled edges. So hard are they that the Indians have used them for carving bone implements. The ends

[2] See La Hontan, Le Beau, and others. In 1795, at London, was published Samuel Hearne's "Journey from Prince of Wales's Fort to the Northern Ocean," which contains a very accurate and interesting account of the beaver. Notwithstanding, we find Heriot and others, as late as 1807, repeating without stint the earlier fables about that animal.

of the incisors by constant use rapidly wear away, and the waste is supplied by a rapid growth from below.

Beaver skeletons have been found, in which one of the incisors was broken off short, while the opposite one had grown so long as apparently to prevent the animal from feeding, and death was supposed to have been caused by starvation.

The tail is flat and oval, nearly a foot long, and covered with a rough, scale-like skin. By the aid of this member the beaver accelerates his speed when under water, or guides his course.

By ten o'clock the heavens gave promise of clearing, and the guides took the two canoes across the carry to Megkwakagamocsis, while Captain Sartor and the writer, with gun and sketch-book, followed, to explore the lake and its neighborhood. It turned out to be a commonplace enough body of water, nearly two miles long, with two deep coves, one on each side of its upper end, and a long stretch of dismal barren, filled with "rampikes,"[3] at its foot. Its water was warm, yellowish, and appeared to contain nothing but large tadpoles.

Paddling down to the outlet, we got out of the canoes upon some logs which had stranded there in a confused mass in the shallow water, when Sartor spied an old beaver-lodge nearby on the bank among the roots of several large cedars. As it was the first one he had ever seen, he and Silas scrambled over the logs and up the bank.

"I say, Silas, how old is this house?" queried the Captain.

"O, I don't know. Mebbe three, four year," replied the Indian.

"Do you suppose there are any beavers in it now?"

---

[3] **2020**: rampike: an erect broken or dead tree.

"O, no. No fresh sign. Beaver all gone last year." And Silas kicked away one or two decaying sticks, the ends of which protruded from the house.

During this colloquy Sartor had picked up a long, sharp-pointed stick, and was vigorously thrusting it between the "timbers" of the deserted lodge, and had succeeded in burying two or three feet of it in the weather-beaten pile.

"There! That must be the kitchen," — turning the outer end around, as he spoke, like the handle of a street-organ. "The cook's not at home, evidently. I wonder where the bedroom is! Upon my word, I believe I feel something soft."

The stick was pulled out, and its end attentively examined by both the Captain and Silas, in search of beaver-hair. Joe and the writer scouted the idea of a beaver being in there all this time. The stick was thrust into the pile again, and in a twinkling there was a heavy swash in the water opposite the lodge, followed by a splash and a scramble among the logs, made by an escaping animal. Joe, who was nearest the place of the commotion, jumped instinctively about four feet over the water to another log, brandishing his paddle and shouting "Beaver! Beaver!" while (candor compels the admission) the writer, with equal precipitation and less judgment, rushed to the canoe for his rifle in such haste as to lose his balance, and for some seconds hung, as it were, suspended over the water, with one hand clutching desperately the end of a dead spruce-bough, and the other resting on the gunwale of his slowly receding canoe. And during all this time the chief authors of the disturbance stood on the bank, the one smiling discreetly and the other fairly shaking with mirth.

*

**BEAVER DAM**

*

We waited quietly for ten minutes, in the hope that we might add beaver-meat to our larder, but our hopes were vain. Silas said this was a "bachelor beaver," which had made the old house his temporary abode, and that there were probably holes in the bank nearby, in one of which he had now hidden himself.

That night around the camp-fire the conversation naturally turned to beavers and their habits, and Joe told us that he had caught beavers that were three and a half feet long from base of tail to end of nose, and that weighed seventy-five pounds; that the average weight, however, was about thirty-five pounds. "The beaver," he said, "stand on his hind legs and walk, when he carry mud for the house, snug up to his neck, like this,"—

folding his arms against his breast. "He carry sticks, too, with his mouth, — sometimes two, three, four— great many — sticks, mouth full, can't carry no more. Then help hold 'em with one paw, same as hand, and walk on other three. When stick in beaver's mouth, end comes over his back. Skin very thick here," pointing to the sides of his neck. "Sticks make'em thick; rub hard.

"Beaver know when winter's comin'. Day before pond freeze over he draw his wood all down under water, so as to have it handy. Next day water freeze up tight. Beaver travel round under ice and work in winter, same as in summer; they make some places, where they use to go, eat, in the night; come out when they kin find hole. We bait traps in winter with fresh branches. Beaver like white-wood, poplar, birch; only eat alder when can't get other kind. Poplar is best, — same as pie for us,"— and Joe's merry laugh at this bit of humor was taken up and echoed by his interested audience.

"Beaver like stay on shore rainy nights, playin' on the bank. He drop little castor, and by'm by another beaver swim along and smell castor, and he must go up and see what is. That's the way trap beaver, —with castor. Put some on stick over trap. Best when wind blows out from shore.

"We kin tell beaver one or two years old by size. After that we can't. Young beaver stay in family until a year old, — sometimes two, three years old. In spring they make house for summer, with sticks, — no mud. Then they look 'round for good chance, and just before winter build good house, with sticks and mud. We goin' have late fall this year. That beaver hasn't built his house yet over on Megkwakagamocsis.

"I ketch young beaver once, O, very small little one. Week after I ketch 'im, I camp on island in Allagaskwigam'ook. Next mornin' beaver gone. I say to my brother, 'Well, I guess our beaver 's gone,' but pretty soon I saw something swimmin' in the water away out, and our little beaver came back. I feed him on gruel— first time. After he was 'bout three weeks old, give him bread, potatoes, — anything. That little beaver he plays like a kitten, and runs around very fast, jumping first with his two front legs, and then with the others. He's very mischievous, and gnaws chair-legs and furniture. One night I put 'im in a box, tight, and next mornin' he had cut a hole in it and got out. He was very clean and neat in his habits, that beaver was. Poor fellow, he went away next spring and got caught in musquash-trap at Squaw Brook."

The ingenuity and skill of the beaver are perhaps nowhere so marked as in the construction of his dams.

These are made of sticks and freshly cut branches, combined with mud and stones, and often attain to a length of several hundred feet, and to a height of from three to six feet opposite the channel of the stream, whose waters they hold in check. At the base these dams are wide, and taper to the crest, which is seldom more than eighteen inches broad, but generally so strong as to bear the weight of a man without being much depressed. The lower sticks and branches in the dam generally have their large ends pointing up stream, a circumstance which may be due to the fact that the beaver in swimming with them holds them near the large end, with the smaller or bushy end thrown over his back. When he reaches running water and stops, the force of the current must tend to swing the bushy end of the branch downstream, in which position it is laid.

The direction of the dams is often convex towards the current of the stream, a form which serves admirably to withstand the pressure of the water above them. This form, however, is by no means invariable, for some times the dam runs concave to the current, sometimes straight across it, and sometimes it is sinuous, or both convex and concave. The tourist occasionally meets with a succession of these dams at short distances apart, the water level above the one reaching to the base of the next higher. However, we may wonder at the beaver's sagacity, and however freely we may accredit to him the possession of a mental principle, we can hardly be expected to allow that he makes these successive dams in order that the resistance of the water below each one may counteract the pressure, to an equal height, of the water above it, and thus serve to strengthen the dam.[4,5] The writer, in going up a stream less than three quarters of a mile, has lifted his canoe over twelve successive dams of this kind, in order to reach the main dam and lodge of the builders. Without going so far out of the realms of probability as to endow the beaver with the wonderful engineering skill suggested above, it seems likely that, as he is essentially a water animal, he prefers to make his journeys in that element, and especially so when in quest of a supply of

---

[4] See "The American Beaver and his Works," p. 99.

[5] **2020**: Referring to the book by: Lewis Henry Morgan. J.B. Lippincott & Company, 1868.

food, which he can thus transport much more easily than by land.

The beaver quickly discovers a change in the level of his pond, and if a breach occurs in the dam it is soon repaired, his work being done generally at night. He is a persistent worker, and frequent breaches, whether made by man or by the elements, are apt to be as frequently repaired.

The beaver is very shy and far-scented, and if, while swimming, he detects the presence of man, he dives, and in so doing strikes his tail against the surface of the water as a signal of alarm. The writer has heard a beaver do this at night a dozen times in succession, when our party was undoubtedly the disturbing cause, but so far away that it seemed impossible that we could have been scented. The noise thus made is very loud, and on a still night may be heard from a great distance.

The Indian, after setting his traps for beaver, takes care to splash with water his tracks and all sticks or ground he may have touched; and yet the writer knows of a case where six people passed and repassed over a dam not seven feet wide, and dragged three canoes over it twice in one day, and during the following night a beaver was caught in a trap previously set there.

The beaver's sight is not as good as his hearing. The writer once fired at one from behind, at less than thirty yards. At the report of the rifle the beaver dove for a moment, and, rising again, swam on unconcernedly. An unusual noise does not appear to alarm them.[6]

They have been known to work by the side of a railroad, along which trains frequently passed day and night. An Indian once told the writer that he had shot five beavers in succession from a tree near their lodge, late in the afternoon, as they appeared one after the other, having evidently come out for their night's work.

---

[6] **2020**: On Moosehead Lake the editor has watched a beaver, in the early morning hours, go about his business stripping a poplar tree of its branches. The said beaver had fell the said tree in order to go about stealing the foliage from our shoreline. He did not seem to mind the video camera capturing his work for evidence of his destruction. There was no traditional beaver lodge locally, but he was storing the branches under neighboring docks.

# Chapter IX

*

Beaver Lodges. — Their Composition and Construction. — Beaver Cuttings. — Camp on Megkwakagamocsis. — High-Low-Jack. — Rest. — Sketching by Moonlight. — Cold Weather. — Joe's English. — Baked Beans. — A Beaver Bog. — Caution.

*

**WHILE** on the subject of beavers, the writer will add the results of some observations made by him in the Maine woods within the past few years, during which time it has been his good fortune to examine a number of beaver lodges and some old "cuttings," and to converse on the subject with both Indian and white trappers.

On the 4th of October, 1882,[1] his Indian guide led the writer to a small pond above Allagash Lake, on a tributary of the stream of that name, where, the previous winter, the former had found what he considered the largest beaver-house he had ever seen. On a small point of land jutting out into the pond, near its outlet, and separated from the mainland by a narrow, but deep, artificial canal, was what appeared to be an immense lodge. It was twelve feet long and six feet wide at the base, rounded at the top, and five and a half feet high at two points three feet apart. The portion between these rounded tops was but slightly

---

[1] **2020**: The year following the trip catalogued in this volume.

lower than the tops, and through it projected three larches,[2] from six to eight inches in diameter, all of which were dead. After half an hour of hard work, a hole was made through one end of the structure, by removing one by one a mass of interlaced sticks, and a quantity of earth, in which they were firmly imbedded to the thickness of about two feet. The interior proved to be filled with water to the depth of six inches, and its sides were covered with mildew. The flooded condition of this interior was due, no doubt, to a permanent raising of the water level of the pond, after the erection of the lodge. Old trappers say this is not an uncommon occurrence. By means of a long pole we found two under-ground exits from the chamber, and a passage-way which ran towards the other end of the structure. This chamber was about four feet in diameter.

An opening, almost vertical, was next made between two of the trees and between the summits of the lodge, disclosing a passage-way about seven inches above water level, which afterwards proved to connect the first chamber with another.

Finally, a third opening was made on the same side as the second, and two feet beyond it. The sticks at this place were

---

[2] **2020**: Larches are conifers in the genus Larix. In the northern United States of *larix laricina*. Larches are among the few deciduous conifers, which are usually evergreen, and lose their needles in autumn. Commonly known as the tamarack, hackmatack, eastern larch, black larch, red larch, or American larch. Indian name *akemantak*. The wood is tough and durable, but also flexible in thin strips, and was used by the Algonquian people for making snowshoes and other products where toughness was required. Could a variation of this genus, larix, be where Dr. Seuss found the name of his character, *The Lorax? (Informationally, the genus Lapsias lorax is a lapsiine jumping spider from Ecuador. It was named in 2010 after the Lorax character.)*

more easily removed than in the other instances, and the interior was soon reached. It proved to be a distinct lodge; and while it connected with the old one, the passage had evidently been made after the latter was deserted, for it left the highest part of the floor quite abruptly and above its natural level, unlike any other exit the writer has ever seen in a beaver lodge. This passage appeared to have been used but little, as across its mouth were two or three sticks which would impede, but not prevent, an exit. In the floor, on the opposite side of the room, and within the latter's limits, was the end of the main passageway, two feet wide, which led directly to the pond, and in which the water was over a foot deep at its very commencement.

This chamber was eighteen inches high in places, and its floor, hard and compact, sloped gradually from the centre to the edge of the water in the passageway, above which in no place was it elevated more, than six or eight inches. The lodge bore signs of recent occupation on one side, and contained fresh grass, and a freshly peeled stick some four inches long. The top of the chamber had numerous pointed sticks protruding through it, but its sides were quite smooth, as likewise was the floor. The interior was five feet wide and nearly round. One of the dead larches helped to support the structure about halfway between the centre and the circumference, and next to the main exit. The beavers had gnawed it about halfway through.

The sticks, which formed a part of the material of these two houses, varied in length from ten inches to ten feet. The shorter ones were generally from two and a half to three inches in diameter, and from this fact, and the presence of these sticks in the walls of the lodge, it seems fair to argue that they were cut short simply for convenience in transportation. A few of these short, thick cuttings were also found in and on top of the dam, which in many places otherwise consisted almost entirely of loam. On the top of the deserted lodge was a freshly cut cedar stick, the only fresh cutting in the place; and during our excavations we also found a few old spruce and larch or hackmatack sticks, from which the bark had never been removed.

The third lodge was more interesting than either of the others. It stood opposite their centre, and its base ran to within two feet of their base, and almost touched the water's edge on two sides. It was very low, and when first seen was taken to be merely a heap of old beaver-sticks brought there for future use. On the side towards the open water of the pond the sticks were almost bare, and were piled up in several irregular layers, just as they had been dragged out of the water, so that it was but the work of a few moments to lift them off and uncover the interior. They formed a wall or roof not more than a foot thick, and there was no mud nor loam on them, except a very little next the chamber, and that appeared to be soil "in place," so to speak. On the other side of this lodge the water of the pond, as before stated, reached almost to the edge of the structure through a narrow and shallow trench, and up this side the beavers had brought a quantity of oozy loam, and put it upon the top of the lodge. In so doing they had scattered it up the entire side for the width of a foot, and had made quite a smooth slide, or pathway, in which could be plainly seen the marks of their feet. The smoothness of this path was apparently due to the dragging of the beavers' tails in their movements to and fro, but careful examination failed to show any distinct tail-marks, or impressions, either here or on the deposits of loam on the top of the lodge, to indicate the use of his tail by the beaver as a trowel.[3]

*

**BEAVER LODGE**

*

The most instructive feature of this lodge was a small hemlock bush, some four feet high, which seemed to grow out of its centre. On reaching the chamber, we found that the stem of this bush protruded into it about two inches below the top. The bush had been gnawed off, and its roots removed as well. There cannot be the slightest doubt that it grew on the spot, and

---

[3] Among the extraordinary habits attributed to the beaver by early explorers, and which have since been denied, this of using the tail like a trowel is the only one which still finds a believer among the more intelligent writers of today. Lewis II. Morgan, in his exhaustive and admirable treatise, "The American Beaver and his Works," on page 29, says: "But he uses his tail to pack and compress mud and earth while constructing a lodge or dam, which he effects by heavy and repeated down strokes." It appears on page 225, that the author rests his conclusion, not on his own observations, but on those of another person. Samuel Hearne was a firm disbeliever in this theory.

this fact, together with others previously noted, leads naturally to conclusions as to the first steps taken in the construction of a lodge. A suitable site is selected by the beavers, within a few feet of the water's edge, and one or more passages are burrowed into it from the water. The hole made above water level, at the surface of the ground, is probably left temporarily small. It is then covered with over lapping sticks, any tree or bush within its limits serving as a prop, but not being necessary for that purpose, during the early stages of construction. When these sticks are several layers deep, mud or loam is added, sometimes mixed with the fibrous roots of grass or water plants, and principally on the highest part of the structure. This mud is so soft that of its own weight, and aided by subsequent rains, it sinks into and fills up most of the interstices between the sticks. Later, as the lodge assumes due thickness and stability, the interior is enlarged, and in some cases the prop is removed. The enlarging process is probably not confined to a lateral extension, but as the top of the chamber may sink from accumulated weight above it, its material is smoothed off, or removed, in the same manner. Late in the autumn the entire lodge is covered with fibrous mud or loam, to the depth of several inches, as a protection against enemies and the cold of winter.

In the new lodge above described the writer found several short and fresh cuttings, one of which, apparently of yellow birch, seven inches long and one inch in diameter, had been peeled, and its fibrous part freshly gnawed along almost its entire length, to the depth of an eighth of an inch.

On our way down Allagash Stream below the lake, we came upon an old beaver clearing on a small island where formerly had stood a grove of mountain ash trees. Many of them — the smaller ones — had been cut down by the beavers and entirely removed, while others lay prostrate, and a few still stood erect, gnawed partly through.

*

**2020 Edition**

**BEAVER STUMP**
**On the shore of Moosehead Lake.**[4]

*

It has been said that the beaver always cuts his trees so that they shall fall into the water, when this is possible. Of the prostrate trunks in this clearing, ten lay landwards, while five lay in or towards the water. The trunks in the water had been stripped of their boughs; the others, as a rule, had not. Three of the largest trees had been gnawed equally on all sides. One of these measured three feet and four inches in circumference at the bottom of the cut, and the next smaller one an inch less. Most of the trees were gnawed almost to the heart on one side, and but little on the opposite side. Many of those which had fallen still clung to their stumps by the shivered tendons of the heart, but in a number of cases both stump and trunk had been gnawed off smoothly, — another evidence, if any be wanted, tending to show that beavers eat of the fibrous wood. Many of the trunks had been partly cut through, three or four feet from the but. It was noticeable that many of the trees were gnawed most on the side away from the water, — and that, too, when their stems grew perfectly straight, — not the side a woodman would cut in order to fell them into the water.[4]

From these observations, and from the further fact, well established, that several beavers often gnaw simultaneously to fell a tree, the inference seems fair that these animals do their tree-cutting from such side or sides as convenience may require or "fancy" dictate.[5]

On the morning after our beaver excursion, when we awoke, there was a thick mist hanging over lake and forest; but it

---

[4] **2020**: A poplar stump left by a beaver who felled this tree to fall perfectly on a rock jetty that extended out into the lake. He proceeded to sit on his hunches on the rocks, night after night, stripping the branches to store for his winter feast.

[5] **2020**: A 2004 study measured the falling angle of 888 beaver-cut trees. They scientifically studied (and found evidence supporting) the hypothesis that the beavers preferentially felled trees toward the location of their dam project. No beavers were available for comment. (Source: Samways, Poulin, & Brigham, "Directional Tree Felling by Beavers," in Northwestern Naturalist, 85:48-52, 2004.)

quickly gave place to a bright sky, the barometer in twelve hours having "risen" nearly half an inch. We packed up our things and were soon on Megkwakagamocsis, on whose eastern shore we pitched our tents, in a spot where the trailing arbutus[6] grew in profusion. From this circumstance we called the place "Arbutus Camp." The afternoon was spent principally in adorning our abode, and in adding a table and some other conveniences to the camp furniture. That day was perhaps the most beautiful one of the thirty-eight days of our absence from home, and was especially noticeable, as it was followed by some of the most execrable weather the writer has ever experienced in the Maine woods.

Towards evening Silas and the writer went to the south cove and set several traps for musquash, and while there distinctly heard a moose "talking," as the Indians say, which consists in making, at short intervals, a sort of subdued grunt expressive of content. We glided stealthily along the shore, and waited long in the hope that the old fellow would show himself, but he did not.

Our favorite amusement, after the day's chores were done, and when we were not too tired and sleepy, was to play a four-handed game of cards. "High-low-jack" is the game, *par excellence*, among the guides, and a great rivalry had sprung up almost from the outset of our journey, between the respective

---

[6] **2020**: Trailing Arbutus (*Epigaea repens*) – also known as Mayflower, or Plymouth Mayflower – in reference to the fact that it was the first flower to cheer the hearts of the Pilgrim Fathers after the rigors of their first New England winter. It grows in a creeping mat, commonly only 4-6 in. high.

occupants of the two canoes, to see which should carry off the championship.

"Well, Silas," the Captain would say, "how is your courage tonight?" "Always good," replies the Indian.

The blankets are now smoothed out in the centre of the tent, a candlestick improvised by sharpening a stick, splitting its upper end, and inserting in the split a loop of birch bark, into which a candle is fitted, and then each player makes himself as comfortable as he can, either by reclining on his elbow, or sitting, Turkish fashion, with a pile of blankets behind his back. The game goes merrily on. Joe deals. "I ask one," says Silas, looking very wise. "Go ahead," says Joe after due reflection, "I give you one — on my partner's hand." Two or three rounds are played. "That's my meat!" cries Joe with a chuckle, being fourth hand, as his partner throws a ten on Silas's ace, and the writer hesitates what to play. The trick is Joe's "meat," for he takes it, and swings his ace of trumps, saying, "Mebbe he's all alone." What shouts of laughter greet the capture of the jack on such an occasion, and how gay is Joe's merry, "Ha, ha, ha! Well, well!" after such an achievement by himself or his partner! How triumphant, too, the announcement by the victors, at the end of play, as to the number of games their opponents "owe" them!

Then comes the final "toasting" of cold feet or hands, before the fragrant and sizzling birchen logs. Our pockets are emptied of their contents, nightcaps drawn on, "heading" arranged, and at last the many tired limbs are stretched beneath the blankets. With a sense of profound thankfulness, we contemplate the flickering of the fire-light on the tent. What quiet! What rest! The smoke eddies and curls in fairy forms, and floats up, — up,

— up, — into Fancy's realms. Thither we follow, led by the silvery cords which entwine us in their meshes. Our limbs, one by one, refuse to act, and finally, enveloped in the incense of the camp fire, we are lost in peaceful slumber.

The next day the traps were visited early, and one muskrat was the only booty found in them; and this was a "kitten," as Silas called it, the offspring of that year. As we held it up before the Captain, and informed that gentleman that there was his dinner, he asked what it was. The question having been answered, he exclaimed jeeringly, "Muskrat! I should call it a musk-mouse;" and ever afterwards, when Silas and the writer went to set the traps, he scoffingly bade us not to bring home any more of those "musk-mice." We noticed, however, that, when the cassambo was made, the dainty bits of fresh meat were appreciated by the Captain fully as well as if they had been large and less tender.

That afternoon the guides, who had been prospecting, reported that they had seen Mooselcuk[7] Mountain and Musquacook Lake,[8] and an intervening bog, over which lay our route, and which seemed to be fine country for game.

We had hoped to call moose that night; but the wind, although very light, was provokingly persistent. The moon was in her second quarter, and, when not obscured by passing shower-clouds, shone brightly, and made radiant with her floods of light the patches of forest tangle. The beauty of the scene was too much for Sartor, who arose and made a sketch of it, after we had all gone to bed. The writer, turning over

[7] "Moose place."

[8] Or Maskωécook, "birch-bark place."

drowsily, noticed the absence of his companion, and looked out to see the cause of it. The Captain sat by the table with paper and pencil before him.

A lighted candle was at his side, surrounded by tin-plates, dippers, and pans, for reflectors and shields from the wind, and these were propped up by forks, boiled potatoes, and a rubber boot. During showers he would dive into the tent, and out again after they had passed, to complete his work. Surely this was sketching under difficulties.

In the morning Silas brought into camp one muskrat, this time a big one. The wind soon began to blow a gale from the northwest, and before night changed around to northeast, bringing with it a blinding snow squall. The mercury went down to 28°, and extra clothing was pulled out from the bottoms of our bags, and drawn on to our shivering bodies. This day and parts of the two following were devoted by the guides to path-cutting and preparing a new camp ground on Bog Brook, and to carrying a part of our luggage up to it. They had found that the old tote-road, on which we were now encamped, ran up the brook for several miles, and over to the head-waters of Musquacook, and, save some needed "bushing out" and bridging, was in very fair condition. For the other members of the party these three days had, unfortunately, a great deal of sameness. Snow fell, accompanied by gusts of bitter wind. At night the sun would go down in a "golden glory," giving promise of fair weather; but in the morning the same old leaden clouds would be there, and the same chilling wind, which froze the water in our pails and almost the marrow in our bones.

One morning while we were at breakfast, the end of our cooking-stick, or kōkwä'took as Joe called it, dropped off and fell into the fire. Joe said this was a sure sign that we should run short of food, an announcement which did not surprise us at all. Joe also said, that among his people the old hunters were very superstitious, and would never allow their kōkwä'took to burn off. He did not seem to think, however, that this precaution was a preventive against hunger; for, added he, "I use to be in the same situation. Now, I look pretty sharp for me. I carry all sorts of grub," with hesitation, "when I kin git it." Joe's English was generally very good, for a self-educated man, but it was often just wrong enough to make listening to him quite interesting. Two peculiarities of it were his hesitation, and the emphasis often placed on the wrong word, especially in cases of antithesis, the latter a peculiarity with which the construction of his own language may have had something to do. Both he and Silas frequently used the present tense of the verb for the past, in accordance with the usage of some Indian tribes.

During our stay at Arbutus Camp, Joe exclaimed one morning, "By thunder, we 're goin' to have a snow — shower;" and when the snow seemed to be coming at last, "We shall have it that time; I know it long time ago; I know it since this morning, when I first git up." But the threatened storm-cloud blew over, after having deposited at the sides of our tents hundreds of little white pellets, which Sartor said made the premises look as "if a homoeopathic pharmacy had been tipped upside down." Two more animals fell into our basket, another of the Captain's "musk-mice," and a young otter (*Lutra canadensis*)[9] that had got caught in one of the musquash-traps, and was very savage as we approached him. "One hide to take home, anyhow," remarked the Captain.

The last night of our stay on Megkwakagamocsis (Harrow Lake) we made a futile attempt to "call" moose, and Joe tried to encourage us by saying that he had heard a moose that day up the brook, and had seen some very large and fresh tracks. "You can't see a bush handy to the stream," he said, "hasn't been cut, — all eat by moose. Small spruce rubbed by bulls. Logging road all nice, — travelled by moose, — all cut up. Moose come in alder-ground now. He's looking for this" — with a gesture of the hand — "this red — willow."

Before we went to bed a good fire was built, over which, at a height of about six feet from the ground, hung a pail of beans. The opinion seems to prevail that for genuine baked beans an iron pot and a bean-hole are essential. We had undertaken too hard a trip to encumber ourselves with a bean-pot, and had made up our minds to eat our beans boiled, but Joe said he could give us as good a dish as we wanted, and that too without either iron pot or hole; and he did it by preparing the beans just as he would for a regular baking, and then hanging them high over a good fire when we went to bed. The writer can recommend beans cooked in this way as a very superior article, and one prepared with comparatively little trouble.

The next day we moved to our new camp, which was near the edge of an extensive bog. Through the bog ran Bog Brook,

---

[9] **2020**: Lutra canadensis river otter European classification, as late as 1964. Currently the species name is Lontra canadensis, and includes all New World river otters.

a stream which for ages probably had been the abode of beavers. Their old dam was at the lower end of the bog, and the sticks had all disappeared from it, leaving a mass of solid earth and gravel, out of which grew several sturdy bushes. Along the brook lay a wide strip of grassy land, which formerly had been overflowed by this dam. A young growth of larches now flourished in it. The stream had worn a new channel through the turf at one end of the dam, the latter having been too solid to be itself swept away. We were very quiet in our movements that afternoon, and no chopping was done, but dry stubs were pushed down and gathered in quantities sufficient for our use overnight. Silas said this was the invariable habit of good hunters, when in the vicinity of game. Joe got a large piece of birch-bark, and from it made a horn, two feet long, five and a half inches in diameter at the large end, and an inch and a half at the other. This was our "moose-horn."

**BEAVER MEADOW ON MEGKWAKAGAMOCSIS**

## Chapter X

*

Moonlight on the Bog. — Moose Calling. — An Answer. — A Big Bull.— The Moose Bird.— Moose and Caribou Tracks. — Antlers. — Beaver Meat. — Birch Bark and its Uses. — A Birchen Age.

*

**AFTER** supper, at six o'clock, the wind having been succeeded by an almost absolute calm, the guides and the writer put on all the clothes they had, which included two pairs of socks and trousers for each, and several coats, and, armed with rifle, tin dipper, and moose-horn, stepped into their canoe with moccasoned feet, and pushed silently off from the shore. Joe, in spite of the writer's misgivings, had previously lighted his pipe, saying that, if he would call the moose up to within two feet of the canoe, he must have a smoke first, "to make it sure." The moon, now in her second quarter, was about an hour high, and from over the eastern line of forest shone resplendent on the bog and stream below, casting dark shadows into the water, and making the spruces and larches look blacker than ever. A gauzy film of mist-cloud overspread the eastern heavens, and a broad ring of light encircling the moon gave token of impending rain. The water was smooth as a mirror, and dotted with lily-pads and long, drooping spears of grass, while over it and near its surface hung a light mist, which floated almost imperceptibly from us towards the head of the bog. The scene was superb, and strangely in contrast with the errand on which we were bent. The silence that enfolded us as we floated onward, impelled by the strokes of the ever-hidden paddle-blade, was broken only by the "Hoo-oo-o!" of an owl, which called forth from Joe in subdued tones the exclamation, "Ugh! We goin' have some rain." The writer occupied the middle of the canoe, his knees resting on a blanket spread over the bottom, while the principal part of his weight was supported by the middle thwart, on which he partially sat. Silas took the stern, and Joe with dipper and horn was in the bow.

Silently and carefully we paddled up the bog, peering anxiously into the blackness under the trees, in a vain search for some moving object. Ever and anon the water before us was rippled like a flash by the retreat of some fish that we had disturbed in the quiet of his shallow home, or we were startled by the plunge of a musquash from his playground on the bank.

When we had reached what Joe considered a favorable point, he laid aside his pipe, noiselessly placed his paddle across the canoe in front of him, and, taking up the horn, dipped its large end several times into the water. Then, putting the small end to his mouth, he called, "Oo-ugh, oo-ugh, oo-ugh, oo-ugh, oo-oo-oo-oo-ugh!" the first notes being short, rounded, and plaintive, while the last one was prolonged several seconds, grew in intensity towards the end, and finished with a downward inflection,— the embodiment of all that is dismal.

The call resounded over the bog, and was taken up by hill and mountain until its last echo died away in the far distance. Hark! What was that? Was it an answering note from out the forest, miles away? No, it was the beating of my heart, magnified a hundred times. And that? Was it the crash of a breaking bough? No, it was the grating of a lily-pad against the side of the canoe. A long and breathless waiting. Our senses are all concentrated into one, that of hearing. A drop of water falls from the idle paddle-blade, and, sparkling in the moonlight, joins its kindred substance with a loud metallic ring. The action of our lungs, in breathing, turns our clothing into bellows, whose discordant creaks must startle any game that comes near us. Hark again! Surely that was no fancy of mine, no deception. Joe heard it too. His head is turned. He listens intently. Hark again! From the depths of the forest far to the north comes a faint sound like the dull thud of an axe, and hardly distinguishable among the many notes Fancy had just been pouring into my ear. "He is coming," whispers Silas. Joe nods his head, and, grasping the horn, calls again, to guide "his Mooseship" to the proper spot. "Oghk!" — I heard it then. "Oghk!" — again out of the wooded aisles, this time more distinctly, and in a deep guttural tone. "Oghk!" — there it is again, nearer still. A tremor comes over my limbs. They quiver with expectation and excitement. Try as I may, I cannot control it. Joe answers again, less loudly than before. The bull is nearing the stream above us. "Let's go up," whispers Joe, and the two paddles dip deeply and silently into the water, as we speed towards our prey. "Oghk! oghk! oghk!" come in quick succession and impatiently from the bank before us. The canoe stops. My trusty rifle is ready and the nervous tremor is gone. Joe's dipper is full of water, and lifted high it pours its contents slowly into the stream, with several pauses. "Oghk! oghk! oghk!" — a mighty crash among the underbrush! The old bull so near his fancied mistress is in a perfect frenzy. Look! look!

See that black mass! The canoe swings slowly around, that I may fire without change of position. What if I should overshoot! Heavens! I cannot see the foresight on my gun against the blackness of the trees. But why does he not come on? He hesitates. He turns. He is coming round, to avoid that muddy brook. We turn again, and slowly paddle up the stream abreast of the calls, which come with frequency from the forest beyond the open bog on our left. Every now and then a dead

*

**2020 Edition**

**MOOSE IN THE BRUSH**
**Springtime when the antlers are growing back.**

*

tree branch is snapped off or broken under foot by the moose in his course, who seems to lay about him violently among the underbrush, knocking against the smaller trees with his horns, and apparently in a state of great excitement. Joe's most defiant challenges and most winning lows fail equally to bring him again to the water's edge. The old fellow is too suspicious.

"Must have cow with him," explained Joe after fifteen minutes of waiting. "Can't get him tonight; mebbe tomorrow mornin'." But the old bull had no cow with him, as subsequent examination of his tracks proved. The few whiffs of tobacco from Joe's pipe had reached the bank of the stream where the moose first came out to it. The odor was strong enough to be detected by his delicate nostrils, and, mistrusting our presence, he had merely circled us to get our wind. The tracks he made were enormous. Did ever disappointed hunter lose any but the mightiest of game, or fisherman any but the largest of trout?

The failure to get this moose was no small disappointment to the writer. He had hoped to take advantage of the presence of his friend and companion on this excursion, to secure a correct drawing of that noble animal, whose extinction, unfortunately, is likely to be a matter of only a few years.[1] Of all the representations of the moose that have heretofore come under the writer's observation, and he has made diligent search for all that could be found, none does justice to the moose's head. Happily, the non-success of our present efforts was amply made up for during the succeeding autumn.

*

*Autumn 1882*

The writer with his Indian guide, Jean Baptiste, had been camping on the shores of Telos Lake,[2] and was on his way to

---

[1] **2020**: Thankfully, this prediction did not hold. Currently, the enemy of the Maine moose is not being over-hunted, but rather the killer known as the winter tick. Since the early 2000s, the moose population in Maine has decreased by approx. ten percent. The current moose population in the state is roughly 60,000 – 76,000 (+/-).

[2] Sahkkahégan, "water connecting with another body of water."

its outlet one morning about nine o'clock, when simultaneously he and his guide were attracted, the one by a noise and the other by the sheen of the sun on the antlers of a big moose. The object of their attention stood at the water's edge on the farther shore of the lake, about a quarter of a mile away, alternately looking up and down the lake, and lowering his head and nose to the ground, as if trying to scent something. At short intervals we could hear his "Oo-ugh!" the noise the writer had heard at first, and we concluded that the old fellow was in search of a mate. Turning the canoe, we began to paddle rapidly towards him. Jean Baptiste once answered the "call," and the moose looked up in our direction; but as we remained motionless, and as the wind was in our favor, he apparently thought his senses had deceived him. The moose's perceptive faculties are reached principally through his nose and ears. His eyes do not seem to have the power of rapid transmission to the brain, at least to that part of it which may be the seat of fear, and neither what he sees nor what he hears has one tithe of the effect on him of what he smells. In this case, although he did not again look directly at us, still he was in a position from which the common red deer (*Cervus virginianus*)[3] would have seen us at once and taken in the situation at a glance.

That the bull was a large one we could plainly see, as he awkwardly went up the slope of a mound and stood half concealed behind a group of birch trees. His long legs, short

---

[3] **2020**: In the 1800s this genus referred to the Common American Deer. (in, “The Viviparous Quadrupeds of North America,” 1848). The red deer (Cervus elaphus) is not native to Maine, but is now farmed in the state. The white-tailed deer (Odocoileus virginianus) is most common to Maine.

neck, massive horns, and black coat gave him a gaunt, ungainly appearance, fascinating from its very uncouthness. Jean Baptiste plied his paddle now to the utmost, and we were rapidly approaching the shore, when the moose moved out from his cover and walked slowly towards the thick woods. The writer's "Winchester" spoke three times: the bull wheeled and faced the opposite direction, holding his head very high. Another shot, and he wheeled again and disappeared behind some spruces. We soon reached the shore, sprang out, and ran up to the edge of the woods, to see our game beyond us in a little opening. He was on his knees; his head and body swayed from side to side, and he soon sank to the ground. We approached him, but he was too far gone even to put his ears back, as hunters say moose do, at sight of man, when they are badly wounded, and in a few moments he was dead. The sun was just appearing above the trees that surrounded this little forest cove, and we heard near by the merry note of the Canada jay, or moose-bird, "What cheer, what cheer, what cheer!" — a most unnecessary inquiry on this occasion, we thought, for he doubtless had been watching our proceedings from the tree-tops, and knew full well what abundant pickings he would have after we should have gone.[4]

---

[4] This bird is one of the tamest that one meets within the woods. So bold is he, that he will almost snatch the meat from a person's fork. No exposed food is safe in his presence, and he ravages the hunters' traps, by picking from them their bait, often to his own ensnarement. His notes are many, the one in the text alternating most often with a "Whŭ—tsch-tsch-tsch!" or with "Whŭee'! [do-sol], Whu-ee'! [do-fa]." By hunters the moose-bird is said to nest and hatch its young in February and March. The origin of another name for the moose-bird is

The writer's camera was at once brought from the canoe, and six plates were exposed before the dead monarch of the Maine forests, and from one of the resulting negatives was made the head of the moose in the frontispiece: the body was drawn from studies of a skeleton in the rooms of the Boston Natural History Society.

Jean Baptiste said our moose was a large one, and would weigh about nine hundred pounds. The following are some measurements we made on the spot. From end of mouffle to line between base of horns, 2 ft. 3 in.; from end of mouffle to top of withers, 4 ft. 7 in.; from end of mouffle to tail, 10 ft. 2 in. The foregoing were measured along the animal's outline, as he lay on his belly with his legs under him, and head stretched out in front with lower jaw resting on the ground. Around the mouffle, about six inches from the end, the measurement was 2 ft. 2 1/2 in.; from end of "bell" to base of horn, 2 ft. 8 in.; girth of neck, half-way from ear to shoulder, 4 ft. 10 in.; length of ear, 11 in.; from tip of one ear to tip of the other, 2 ft. 7 in.; from hock to tip of hoof, 2 ft. 7 in.; from hock to tail, 2 ft. 9 in.; from middle of duclaw to tip of hoof, 9 in.; width of antlers at tip, 2ft .8 in.; width at widest part, 4 ft. 2 in.; circumference of horn at smallest part near base, 8 1/2 in.; weight of horns and dry skull,

---

thus accounted for by Dr. J. Hammond Trumbull, in a note in the Collections of the Connecticut Historical Society: — "'Whiskey Jack,' the name by which the Canada Jay (*Perisoreus canadensis*) is best known to the lumbermen and hunters of Maine and Canada, is the Montagnais Ouishcatcha$^{n}$ (Cree, Ouiskeshauneesh), which has passed perhaps through the transitional forms of 'Ouiske Jean ' and 'Whiskey Johnny.'"— Vol. II. p. 44.

**2020**: Moose-bird was the name for the Gray Jay, now named the Canada Jay.

39 pounds. The writer may be pardoned for adding that the head and two hind quarters were at once sent to Moosehead Lake, and the greater part of the fore quarters was afterwards used at Chamberlain Farm, so that very little of the meat was wasted.

*

The first question asked by the novice in the woods when he sees a large track, and is told carelessly by his guide, as Joe once expressed himself, "Mebbe moose track; mebbe caribou track." How can one distinguish between the two? The track of the cow-moose is more pointed than that of the bull, while both are much more pointed than that of the caribou. This is the chief difference. Again, in the caribou's hoof the outer edges are sharp, and the bottom of the hoof slopes upward from the outer to the inner edge. The bottom of the moose's hoof is fuller and flatter. Consequently, the caribou's track in the mud is rounder, and deeper at the sides than in the middle, while in the case of the moose it is long, pointed, and almost equally deep everywhere.

**CALLING MOOSE**

It seems to be pretty generally admitted that the age of the moose can be correctly estimated by the size and appearance of his horns, although this is not invariably the case. A moose in good condition, strong and vigorous, will have thicker and heavier horns than one of equal or greater age that happens to be sickly. The time of shedding their horns also varies, but as a rule this takes place early in December. An Indian hunter of experience once told the writer that he had killed two moose together in the first week of February. One was "poor," and still wore his horns, while his mate was much larger, and had lost his, and had new ones growing, an inch long. With caribou, too, there seem to be exceptions which would indicate that the physical condition of the animal has much to do with the shedding of its horns. The same Indian mentioned above, and a white hunter also, said they had each shot a caribou on the 20th of November, in different years, and both animals had lost their horns. In one case the skull where the horn had joined it was still red, while in the other it had already healed over and become smooth. Both were fine-looking animals. The same two hunters said they had found bucks, seven or eight years old, wearing their horns in March, — once as late as the 20th of that month.

When the horns of the deer family are young, or "in the velvet," — say from April to August, — they are very soft, and bend readily, but will rarely break. They are also very tender to the touch, and at such times the animals seldom venture into dense underbrush, but frequent open bogs and meadows. The female caribou sometimes has short horns.

The day after our unsuccessful moose hunt, the guides went on their usual exploring expedition, and returned at seven o'clock, after dark, tired and wet. They said they had been over on Fourth and Fifth Musquacook Lakes, having gone up the bog and past Clear Pond;[5] that their way had led through cedar swamps and mud-holes, sometimes taking them knee-deep into slush and water. Very luckily, they came out of the woods opposite camp, having walked some distance in the dark, guided, as they said, only by the wind. Joe had set two large traps on their way up the bog, and early the next morning brought into camp our first beaver. The Indians are very fond of the flesh of the beaver, and the arrival of this one was hailed with great satisfaction, Silas remarking, as we afterwards sat at table picking its bones, "That goes right in the spot."

Beaver meat has a decided flavor, peculiar to itself, and often slightly disagreeable at first to one who never has eaten it. This flavor is more noticeable as the animal is older, when the flesh needs considerable "bo-boiling," as Joe used to call it, and a generous seasoning of pepper. The tail is esteemed a great delicacy, and is eaten roasted. It is fleshless, and of a soft, gelatinous nature. The talk among guides about "beaver-tail soup" seems to be, so far as the writer can learn, a bit of pleasantry, and nothing more. The liver and the flesh next the tail are considered the best parts of the animal. The flesh is always eaten boiled, while the liver is spitted on a green stick, roasted, and served up with butter, salt, and pepper. It is a most delicious dish.

---

[5] Nuka$^{n}$congamoc, " head-water pond."

Our next care was to stretch and dry the beaver's skin. The stem of a long cedar bush was cut, trimmed, bent into the shape of a hoop, and the ends bound together with cedar bark. Small holes were made an inch or more apart around the edge of the skin, and through these pieces of cedar bark were passed, and tied over the periphery of the hoop. The latter was considerably larger than the width or length of the skin, which could thus be stretched at will. The hoop was then hung up in the open air where it would not be exposed too much to the sun, and in a few days the skin was dry enough to pack away in a bag.

The Captain was very much elated that *his* guide had brought into camp the first "respectable" supply of meat, as he termed it, and expressed a desire to honor the occasion in some unusual and suitable manner; but as the writer had the only flask of spirits in the party, and as the Captain was, besides, morally opposed to the use of liquor, he reluctantly fell back on a bottle of acid phosphate,[6] which Joe modestly declined, with thanks. A matter of no little consequence to the Captain was the fact that his tobacco was giving out. He had smoked recklessly thus far, and his supply was now reduced to a thin, narrow plug, about five inches long. On the supposition that we should be seven days more in getting down to Allagash Falls, he divided this plug into seven imaginary pieces, and said he would limit himself to one a day; that when it was all gone he would dispose of his pipe, and he forthwith tried to make a dicker for the latter with Joe, in advance. But Joe, supposing him to be in earnest,

---

[6] **2020**: A drink with a pH between 2.0 and 2.2, or about the same as fresh squeezed lime juice. Also used in the late 1800s and early 1900s to make soda fountain drinks.

generously said he didn't "want buy a pipe out of a man, and he go without a *pipe*" and so the dicker ended. As a matter of fact, the Captain's seven days' rations lasted just a day and a half; but it is due to his foresight and prudence to say that he had previously learned from the guides that they had more tobacco than they would need, and Joe had said, "When we get short, we 'll divide the balance."

Our vicinity to Musquacook, the land of the birch, had naturally led us to inquire about the uses of birch, and especially of the bark of that tree; and among other things, Silas said he had heard of Indians boiling green corn in the ashes in a birch-bark vessel, and Joe said he had seen water boiled in that way. As the day was cold and raw, Sartor and the writer decided to stay in camp, and to try experiments. After having replenished the fire, we cut several pieces of bark from some neighboring white-birches, and each made a small "bucket," by overlapping the corners and pinning the folds to the ends of the vessel with a small bit of a twig. The buckets, filled with water, were then placed upon a bed of glowing coals, and we awaited the result with much incredulity as to our success. We first used rather thick bark, and when exposed to the flames it burnt on the outside and down to the water's edge. The principal difficulty with any bark, thick or thin, aside from cracking "eyes," was that the holes through which the fastenings or pins were put leaked, and when the water-line got below them, the bark burnt, and then a collapse was imminent. This actually happened, in part at least, to our first buckets, but not until the water in them had become hot enough to scald our fingers.

We now took a piece of very thin and pliable bark, and made a bucket six inches long, four inches wide, and four inches high, and poured into it nearly three pints of water. A bed of fresh glowing coals was raked together, the bucket set on them, and then hot coals and fire-brands were put around and near it. This bucket was a success in every particular, and the writer has preserved to this day the bark which formed it. Apart from having a few black scars, it is as good as ever. When the water began to simmer, we put into it an egg. As our incredulity was still great, we left the egg in ten minutes. It would have done credit to an Easter festival. Then we tried another for five minutes, and it was very well timed. If the water had then boiled as it did later, three and a half minutes would have been, beyond doubt, long enough. Next a small potato was put in, and in twenty minutes it was thoroughly boiled. Finally, a handful of pearled barley was tried, and in thirty minutes it too was cooked. There was left in the bucket one pint and a quarter of water. The fact, then, is established, that for iron pots and pails birch bark may be used as a substitute, and gentlemen campers who lose their "kits" in the rapids, if they are fortunate enough to save the potatoes, need not despair of having these boiled for their next meal.

Of the various uses to which birch bark may be put in the woods, the first to suggest itself, probably, is that for canoes. Next in importance is its use for kindling fires. It is also, but less frequently, used to make wigwams or "shelters," canisters for molasses, buckets, dippers, candlesticks, torches, and, again, as a substitute for wrapping-paper, or to put under small game when being skinned or cut up; and lastly, as we have already seen, from it are made "moose-horns."

Should the human race ever come to a "wooden age," and the iron in our axe and knife be replaced by birch, for example, we shall see the Indian coursing down streams in his birch canoe, or impelling it over lakes with his birchen paddle; "calling" his moose with a birch-bark horn, shooting him with a birchen arrow, skinning him with a birchen knife, cutting birch logs with a birchen axe, kindling his fire with birch bark, boiling his birch partridge in a birch-bark pail, eating him from a birch-bark plate with a birchen fork, drinking his birch-twig tea from a birch-bark cup, wiping his mouth with a birch-bark napkin, ornamenting his squaw's dress with birch-bark silhouettes, bringing up his pappooses with a birchen switch, and, finally, going to bed in a birch-bark wigwam by the light of a birch-bark torch. In fact, his whole life will be birchen, from being rocked in a birchen cradle to being buried in a birch forest.

A few desultory snow-squalls during the day were followed at night by a clear sky and a crisp, cold atmosphere. The barometer was rising, and with it our spirits at the thought of leaving this dismal camp. Two flocks of geese flew over us at evening, which Joe declared to be a sign of coming cold. This tallied also with the saying of an old Indian hunter, that it was “time to quit hunt; de wild goose, *he* come.”

# Chapter XI

*

On to Musquacook. — Mink. — Mud Lake. — The Guides leave Camp in Quest of Food.— Spectral Light.— Apprehensions. — Return of the Guides.— Low Water. — Shoeing the Canoes. — Snow. — Insubordination. — Black Cat.

*

**THE** next morning the mercury was at 20°, and almost the entire surface of the brook was covered with a film of ice, which quickly melted, however, under the influence of the sun's rays. Cold nights were not welcome to us now, for we needed full streams to carry us out of these remote places, and the frost, as Silas expressed it, "cramped" the water too much.

We broke camp early, and struck out for Musquacook. The writer went ahead of the other members of the party, and, having reached the lake and laid down his pack and gun at the side of the carry, was standing quietly on the shore, when a little mink (*Putorius vison*)[1] ran along the water's edge and approached him. The little fellow did not know what to make of the intruder, and stood repeatedly on his hind legs and sniffed the air with his little pointed nose, trying to make out what that great creature could be. Then he popped under a log, and, coming out at its farther end, began to investigate the pack.

---

[1] **2020**: Early taxonomy classification when American and European mink where in the same genus. The American mink has been reclassified to its own genus and species (Neovison vison).

Several times he made two little hops, or spring-like movements of the fore legs, from side to side, while the hind legs remained fixed in their place. His investigations seemed to be satisfactory, for he soon darted off and did not reappear.

We pitched our tents on a little sandy point on the west shore of Fourth Lake, near where the line between Penobscot and Aroostook Counties was supposed to run; but diligent search on several occasions failed to find the line. That night we saw a brilliant display of Aurora borealis. At this place the guides made a kitchi′pläkwä′gn, which consists of a frame of two upright crotched posts, one on each side of the fireplace, with a horizontal pole laid over them. From the cross-pole depend two or more twisted withes, each ending in a crook, on which are hung kettle and pail for culinary purposes.

*

**NORTHERN LIGHTS ON MUSQUACOOK.**

*

The next day, Sartor, Joe, and the writer visited Fifth, or Mud Lake, which was just above us. We picked some cranberries from a bog on its east side, and saw two otters and gave chase. They were too quick for us, however, and, after swimming under water a long distance, would rise to the surface for an instant to take breath, and go down again with the rapidity of a loon. They finally reached shore, and ran up into the woods and out of sight. About half of this lake was so shallow that we could with difficulty paddle through it. There were several beaver-houses around it, and on the mud beneath us we could frequently see the marks made by the beavers' tails, where their owners had passed along.

The next morning Joe and Silas, with enough bread in their pockets for one meal, took Joe's canoe, and left for Depot Farm on the Allagash, twenty miles away, in quest of food. They knew there was a road which led to the farm from the foot of the Musquacook chain of lakes, but what kind of a road it was, whether of the Mud Pond Carry order or otherwise, and what obstacles they should encounter before reaching it, were matters of which they were profoundly ignorant. They also took with them the Captain's revolver, and two traps to set for beaver on the way. We bade them Godspeed, and after their departure amused ourselves as best we could. Although we had a generous woodpile, we did not care to draw on it too extravagantly, and killed time by cutting more firewood and bringing it into camp. We estimated that, with no mishap, the guides should be back on the afternoon of the second day; and in order to satisfy ourselves of our bodily welfare in case they should be delayed beyond that time, we took an inventory of our provisions, which we found to consist of the following articles, to wit: a pint and a half of pearled barley, two eggs, two cans of tomatoes, two small cans of soup-powders, a handful of beans, a piece of bacon little larger than one's fist, four slices of pork, a pound of dried peaches, a pound of canned beef, and a suspicion of tea, coffee, and sugar. Priceless treasures these were indeed, and carefully we hoarded them.

The first day passed away pleasantly enough, considering our enforced confinement. That night, as we looked up the lake, on the opposite shore there gleamed a bright light for a few moments, like the flame of a candle. It twinkled in the dark just above the water's edge, and, as we watched it in silence and wonderment, it went out. The guides, on the day when they had first visited this lake, had reported a fire among the trees in that very place; and although at the time of our arrival it had apparently gone out, our weird and phantom light was doubtless caused by the blazing up of some thitherto dormant ember.

Morning came and with it more leaden skies and raw winds. By four o'clock in the afternoon, however, the clouds had entirely disappeared, and the wind went round into the northwest. It was growing bitterly cold, and, aside from our own prospects for the night, we began to speculate upon those of the guides, and upon their whereabouts. Anxiously we awaited their return. At times we thought we heard them shout or whistle, and again and again did we go down to the shore, and look towards the outlet of the pond. Hour after hour went by, and they did not come. "What could have happened to them! What if they should not return the next morning! Something must be done. It would take us at least three days to reach the Allagash settlements, if not four, and our food, even with the greatest economy, would barely last so long. It really was not more than enough to last two hungry men two days. Leave we must the next day, at all hazards. With this determination we ate the two remaining eggs and a dipper of soup made from some of the soup-powder, and went to bed. The cold was intense, and we were glad to appropriate, for additional covering, the blankets left behind by the guides.

Morning brought no solution of our difficulties; so, with apprehensive minds we took down one of the tents, gathered our personal effects together, and with blankets, axe, a few necessary dishes, and our scant store of food, loaded our canoe, and at ten o'clock paddled away towards the outlet. We had just reached that point when to our great relief the guides appeared. The writer says relief, for although it was a bitter disappointment to have to return to the least attractive campground we had had since we left home, we nevertheless felt much satisfaction at having our party once more together, — a feeling that was greatly increased afterwards when Sartor and the writer realized more fully what we should have had to undergo had we continued on our way alone. The guides reported the water very low, said they had had to carry their empty canoe up the stream between Second and Third, and Third and Fourth Lakes, and that we couldn't possibly get down to the Allagash without "shoeing" the canoes. They complained that they were wet and almost frozen, and urged us to return to camp, and not to think of going on without them. So turn about we did, and in the face of a cold stiff breeze from the south east paddled slowly and reluctantly back. Silas had bought at Depot Farm three pecks of potatoes, twenty pounds of flour, and some saleratus,[2] half of which he had with him, and the other half he

---

[2] **2020**: A leavening agent consisting of potassium or sodium bicarbonate.

had left in a hunter's camp at the foot of Second Lake; so, after having reinstated ourselves in the old spot, the first thing we did was to have what is vulgarly called a "square meal," which in this instance consisted of corn-beef hash and bread *ad libitum*, and O how good it was!

The guides had reached the Depot Farm the same day they had left us, and had started out on their return the next morning early; but after having gone some distance they discovered that Joe had left his axe five miles behind him, and he had to go back for it. They also said they had shot at and wounded a caribou with the Captain's revolver, but whether subsequently chasing the caribou constituted going back for the axe, or not, the writer has never been able to satisfy himself. At any rate they had spent the night in the aforesaid hunter's cabin, where they breakfasted the next morning on saleratus bread, baked in the ashes, and potatoes without salt.

The work of the day now was to shoe the canoes. The first step was to find several cedar trees whose trunks were straight and free from branches for twelve feet from the ground. Then the trees were felled and their trunks divided into strips nearly eleven feet long, three or four inches wide, and an inch or more thick. These strips were neatly squared, or trimmed, with axe and "crooked knife," and then split into pieces a third of an inch thick, ten feet six inches long, and tapering from three inches wide at one end to an inch and a half at the other.

The "crooked knife" is a very important and often very serviceable part of a hunter's outfit in the Maine woods. Its blade is three or four inches long, straight, or slightly curved at the end, narrow and thin, while the handle runs straight out from it to the extent of the width of a person's hand, and then turns upward at a right angle in the plane of the blade and towards its blunt edge or back. The operator holds the knife, edge towards him, his thumb resting flat against the projection of the handle along its entire length. In cutting he draws the knife towards him.

With our two tools, knife and axe, the work went slowly on, especially as only two of the party could work at a time. Fortunately, cedar is a very easy wood to split, and very few of the strips were spoiled in the process. By night the pieces, fifty in number, were all cut, and half a day's work remained, Joe stating that by "tomorrow evening, proba'bly, we 'll get a little down further." During the night it rained hard for a few hours, much to our satisfaction. Four inches more of water in the streams would carry us comfortably to the Allagash.

In the morning work was resumed by eight willing hands, for now the hunting-knives and larger pocket-knives in the party were available. The splits were all marked off with a lead pencil into lengths equal to the respective distances between the several thwarts of the canoes. Opposite these marks diagonal slits were made, two inches wide, which entered the middle of each edge, and came out, nearly together, on the flat side of the split. Where the splits were narrow, these slits, instead of being diagonal, could without danger of breakage pass directly through them, from edge to edge.

The next step in the process was to string the splits together. Long ribbon-like pieces of the inner bark of the cedar, nearly two inches wide, were already at hand. The slits were forced open with wooden wedges, and the ribbons inserted and drawn through them successively until twelve or thirteen splits were side by side, and the necessary width had been attained. The ribbons projected two feet or more from the sides of the "shoes," which were now declared ready to be put on. Each canoe has two shoes, and, it should be added, the wide ends of the latter, where they overlap, have their surfaces bevelled and counter-bevelled, that operation having been accomplished in the early stages of the process. The shoes are now laid on the canoes, the bow shoe of course lapping over the stern shoe, and the ends of the ribbons are split, carried up the sides, and tied securely each to its corresponding thwart. The narrow ends of the splits are notched instead of being slit, and tied together with cord or cedar withe, which is in turn carried over the bow and stern respectively, and there fastened. The bottom of the craft is now smooth, and without projections, the exposed parts of the ribbons being inward next to the bark of the canoe, and the *voyageur* can go pell-mell down a shallow, gravelly stream without that constant dread of "scraping," which otherwise is so harrowing to the mind.

The work on the canoes took rather longer than we had expected, and before it was quite finished, in the middle of the afternoon, the clouds, which had been gathering all day, poured down upon us a deluge of snow that continued all night. In the morning the snow was ten inches deep, and with rain and sleet was still falling at short intervals. Nevertheless, go we must. Tea and coffee would give out in another day; our last potatoes and flour we had had for breakfast, and only half a mess of beans, a can of tomatoes, a little soup-powder, and some dried peaches remained.

The rain and snow had done one good thing for us, and that was to raise the level of the lake some three inches. We accordingly found the stream below it almost uniformly high enough to float our loaded canoes comfortably. Now and then, seeing a gravel-bar ahead of us, by vigorous pushing we got under good headway and slid in triumph over it. Some little wading was necessary, and we had to cut out a few obstructions, but on the whole, we made good progress. At the head of Third Lake we found a beaver, caught in one of the traps which Joe had reset on the way back from Depot Farm. This was a godsend to us, as afterwards appeared.

At half-past twelve o'clock we reached the foot of Second Lake, and lunched in the little log camp where the guides had recently passed a night. The potatoes and flour left there were found intact, and a fresh loaf of bread was soon baking before a hot fire, in front of which we all warmed our cold limbs. It was here that occurred the only trouble we had with either of the guides, insubordination as it were, which however soon subsided. Sartor and the writer had for many days openly deplored our enforced confinement in the woods during the abominable weather we had been having, and we were also anxious to reach home for business reasons. When, therefore, at one o'clock, Joe said, "Well, I guess we 'll stop here tonight," we pleasantly, but with firmness, declined to do so. This caused the first exhibition of temper on Joe's part that we had seen, and after lunch, when he went out to pitch his canoe, some unofficial parleys were held through Silas, a self-appointed mediator, the outcome of which was that he and the writer left at once for the foot of First Lake and were shortly followed by the others. No reference was afterward made to this little episode, and Joe's general affability and readiness to please made us forget a momentary outburst of feeling on his part, which he himself doubtless regretted.

*

**HUNTER'S CABIN**

*

Near the foot of First Lake our attention was drawn to the shore, on which something was jumping about in the snow. On approaching nearer, we saw a rabbit hop through an open space for some distance in a straight course, and then hop off at right angles into the bushes. In a moment a "black-cat," or fisher (*Mustela canadensis*),[3] came hopping along too, in the rabbit's tracks, and seemed to follow the trail with the greatest ease. We had not long to wait before Bunny came out again where he had

---

[3] **2020**: Members of the Mustelidae family (weasels, otters, wolverines, badgers, etc.). Closest relative is the smaller American marten. Voracious carnivores, fishers will also eat nuts, seeds, berries, and mushrooms. They are one of the only species known to kill porcupines. *Mustela canadensis* is early genus. (Ref. in, "The Viviparous Quadrupeds of North America," 1848.) Current classification is Martes pennanti.

first appeared, and went his usual round leisurely, seeming little disturbed by the fact that his pursuer wanted him for supper. Then again came the "black-cat," his long bushy tail dragging after him in the snow. We had now floated so near that he scented us and scampered away, and the rabbit, no doubt, owed his preservation, for that day at least, to our presence. The fisher, except in color looks at a distance not unlike a fox, and besides rabbits he eats fish, as his name implies.[4] Silas said the fishers built their nests high up in the trunks of hollow trees.

---

[4] **2020**: It is now understood that the fisher is an opportunistic hunter and may eat dead fish found along a shore, but they do not fish. The name *fisher* may have originated from 'fichet' a French word referring to the pelt of a European polecat.

*

# Chapter XII

*

Among the Boulders. — Spirit of the Rapids. — Camping after Dark. — The Allagash. — Twelve Miles or No Supper. — Forest Fires. — Moirs. — Allagash Falls. — Tow Boats. — St. John River. — Sights along the Way. — Retrospect.

*

**WHEN** we reached the dam at the foot of First Lake, nothing was said by either of the guides about camping, and, as there seemed to be no good spot for that purpose nearby, we transferred our canoes to the other side and began to go down the stream. In a very few moments we were in the midst of huge boulders, among which in many places it was impossible to find a channel, and an attempt to wade was generally followed by a plunge from some slippery rock up to our boot tops in deep water, or by the wrenching of an ankle as the foot got jammed into some narrow crevice of the rocks. And now we saw the utility of canoe-shoes, for when our further passage was blocked by some low, flat-topped boulder, we simply pushed the canoe up on the latter, and then dragged it along and over into the water again. Meantime daylight was slowly fading away, and the forest on both sides of the narrow stream shut out much of the little light that still remained. The waters boiled around us, and as we paused now and then in the midst of some difficult "pitch," out of the greater din and roar came a solemn monotone, "Swōsh, swōsh', sŏsh, sŏsh, — swōsh', sŏsh, sŏsh', sŏsh,— swōsh', sŏsh." It was the Spirit of the rapids chanting ceaselessly her soft refrain, whose measured vibrations stole gently into our inner natures and held them spellbound.

Lulled by these pleasant fancies into forgetfulness of the outer world, we were rudely awakened to the fact that night was upon us. Thus far along the stream there had been no possibility of camping. The banks had been steep and densely wooded. Fortunately, we had now reached an old "landing," on the top of which was an even surface, and to it, at a quarter to six o'clock, through the snow and bushes, we dragged our heavy loads. If anyone thinks there is pleasure in pitching camp after dark in a strange spot, with a foot of snow on the ground, just let him try it. In the first place the snow had to be — yes, kicked away from the site chosen for the tent. Then came the search for tent-poles and pins, and for firewood, all of which had to be gathered by means more of feeling than of seeing. No boughs could be had, and the thought of lying on such a wet hard place as we had been compelled to choose was anything but pleasant. There was not, however, a word of complaint heard from any one, and when the fire had once begun to blaze and crackle, and the rubber blankets had covered up the dampness, and our beds were spread out, comfort prevailed again.

It was late the next morning when we continued on our way down stream, owing to a delay of an hour in mending a strip in one of the canoe-shoes. Silas's rubber boots had come to grief, too. While crossing the carry to Musquacook he had stepped upon a stub, and made a hole in one boot, which he had subsequently mended by filling with resin. The writer had in the same way repaired a tear in one of his boots, but it remained water-tight, while Silas's did not. The latter's were accordingly laid aside, and Silas, drawing from one of the bags his leather “driving” boots, which had become dry and hard, threw them into the water for a moment to “grease” them, as he said, before putting them on.

Sartor and the writer, as usual, walked ahead of the canoes, wading from one side of the stream to the other, according as either shore offered the better footing. The stream was rapid, as a rule, but grew more navigable as we proceeded, and in places was even narrow and deep. At one such spot we built a fire just before noon, and sat silent spectators of the still life around us, musing until the canoes should come. A spreading cedar covered us with its branches, from which, under the influence of the ascending heat, water dripped continuously, accompanied by an occasional dab of melting snow. Opposite us, over the water's edge, bobbed nervously the leafless stem of a slender bush, whose half-drowned offshoot quivered unceasingly in the current of the brook. Now and then there would sail out mysteriously, from behind a root or alder-bough in the dead-water, a phantom-like accumulation of foam from the rapids above, looking like a fairy iceberg, and impelled, as it were, by some breath of air too light to be perceived. Then, in contrast, came Joe poling silently along, clad in newly-made zouave costume. One leg of his trousers had become badly tattered below the knee, and, the other being little better, he had cut both off at the knee, and now wore them outside his blue overalls, with his gray socks drawn up over the bottom of the latter. Finally came Silas, perched upon the top of his load, in mid-canoe, balanced by his setting-pole, and looking like a tight-rope walker.

Our course down Musquacook was without further interest. Destructive fires had run through the forest for miles on each side, and had left a picture of desolation. We camped once more on the bank of the stream, in an oasis in the blackened waste, and reached the Allagash the next day at noon. Here we had for lunch the tail of our last beaver, the last can of tomatoes, and the remainder of the soup-powder. Not a mouthful of food was left, and the alternative now before us was a twelve-mile paddle or — no supper.

*

**2020 Edition**

**GOD'S COUNTRY SIGN – KOKADJO, MAINE**
The sign was erected in the 1930s by the Maine Forestry Service and the Civilian Conservation Corps., as a message to be careful with fires.

*

*

**THE SOCIAL HOUR**

*

The forests along the Allagash were no more attractive than those of the Musquacook. Fire had run through them also, and on this bleak October day the charred trunks were in mournful contrast with their white background of snow. All of this ruin, which involved probably millions of dollars, is supposed to have originated from the carelessness of two hunters during the preceding summer, who are said to have left their campfire burning behind them. Too much cannot be said about the importance of putting out fires when a camp is left. The writer has adverted to this subject in print elsewhere, but on account of its importance he may be pardoned for reintroducing it here. A live ember left in the turf at the side of one's fireplace in the woods may be the germ of a terrible conflagration, like the one just referred to, when settlers on the Allagash had their household effects out on the riverbank for a week, ready to leave their threatened habitations at a moment's notice. It behooves all to take the greatest possible care to prevent such a catastrophe. Five minutes spent by willing hands can with little labor make sure of safety in this respect.

We needed no urging to do our best that afternoon, but the weary miles went all too slowly by, and the raw winds chilled us through and through. At each bend in the river our eyes would pierce the misty, frozen atmosphere, seeking the welcome smoke from some settler's chimney, only to be disappointed again and again. Finally, as the shades of night began to close about us, we reached the house of Finley McLennan, on the left bank of the stream, and all huddled about the glowing stove in that farmer's kitchen. On our making some inquiries as to whether we might sleep in his barn overnight, McLennan told us, with a knowing look, that he didn't keep "public house," and said we could get "something" at Moir's, a mile below, where we could also put up for the night. The

Captain promptly repelled the insinuation that we had any greater thirst than could be quenched by a glass of milk; and while it was still light enough to see, we hurried on to Moir's. There we were welcomed cordially, and were soon comfortably housed in a human habitation, — the first time in a month.

That evening was spent in profitable conversation. Mr. Moir, being an old settler on the Allagash, was very well informed about that section of the country, and talked well and intelligently. He said that a good many "sporters" had passed by during the summer, and often stopped at his place for milk and eggs. He had a large family of children about him, several of whom were married to neighboring farmers, while the youngest slept quietly in a cradle near us as we talked, the light from a kerosene lamp shining full in the little fellow's face. When bedtime came the old couple insisted upon giving up to Sartor and the writer *their* bed, on which the favored guests stretched themselves and were soon fast asleep. But our slumbers did not last long, for the unusual heat from the large stove, which was replenished from time to time during the night by the good housewife, baby in arms and pipe in mouth, made us restless and wakeful.

The next day we ran down the Allagash and "carried" past the falls, where the river breaks over ragged walls of clay slate and leaps madly into a pool below. Joe thought the height of the falls was about seventy feet, and said he had heard that measurement given by others, better informed than himself. Moir had given us the same estimate, but was somewhat shaken in his belief of its correctness, because two men, who had worked near the falls for three weeks during the preceding spring, had told him their height was at least ninety feet! We measured the perpendicular fall of the water, first with our aneroid,[1] which made it, as nearly as could be read, twenty-five feet, and afterwards by means of a string, which made it twenty-seven feet![2]

On our way down the river we passed several flatboats, or scows, drawn by horses, and loaded with hay and supplies for the loggers who were to operate up the river during the ensuing winter. Most of the carrying-trade on the Allagash and St. John above railroad connection, at least for the loggers, is done by means of these flat-bottomed boats, which seldom draw, even when loaded, more than eight or ten inches of water, and carry twelve and a half tons' weight, or one hundred and twenty-five barrels of two hundred pounds each. The horses, which weigh from thirteen to fifteen hundred weight, walk tandem, two to each boat, along the river bank, on tow-paths where there are any, and, in the absence of these, in the water, their drivers in the latter case often following them, even up to the waist. This comes to be a severe strain on the men, as well as on the poor beasts, when the snow and slush of an early autumn freeze about their legs. The harnesses often freeze, too, and are never removed from the horses. The latter often have to swim from five to seven rods through stretches of deep water. They are whipped up before they reach these places, the towrope is drawn in, and when the horses begin to swim, the rope is

---

[1] **2020**: An instrument that uses pressures to measure change in height. The 1844 invention of the aneroid barometer is credited to French scientist Lucien Vidi. Now you can use an app on your smartphone.

[2] **2020**: This is a fairly accurate measurement of the series of drops.

slackened, and the headway gained by the boat enables them to reach a footing before they must begin to pull again. The drivers at these times stand on the horses' backs. For these services drivers get from fifteen to twenty dollars per month and their food, which consists of tea, bread, codfish, pork, beans, and dried apples. In former years they were not permitted to stop and build a fire for meals, because that took too much time. They have three meals, except when on the "drive," between the first of March and the end of the drive. Then they have four meals daily, — at four and nine o'clock in the morning, at one or two o'clock in the afternoon, and at eight in the evening. They put their tea and molasses into the same pot, and boil them together.

We reached the junction of the Allagash and St. John early in the afternoon, and at "Negro" Brook Rapids,[3] a half-mile below it, we went a shore and "cast" our canoe-shoes. They had done us excellent service, not only on Musquacook Stream, but on the Allagash as well, which in many places was quite shallow. The scenery from this point, looking up stream, is very fine. Distant mountain ridges rise one behind another, covered with a misty blue, while above the broad expanse of river, and between it and the background of mountains, hangs a strip of grass-tufted meadow, out of which reach upward a few graceful

---

[3] **2020**: This is the one word that has been changed in this text, where the original contains a more offensive term. When the Maine legislature passed, "An Act Concerning Offensive Names," Maine brooks, rapids, ponds, and lakes in Aroostook County, as well as other locations in Maine, were rightfully renamed. (Source: Legislative Document, H.P. 1712., January 10, 2000.) This location might be referring to what is now Casey Rapids.

old elms. A group of mouldy buildings, shaded by a clump of mingled spruce and birches, stands on one side, and here and there scattered about, dotting the landscape, may be seen a ruined mill or dilapidated shanty. The tinkle of a distant cowbell draws the eye to where a group of cattle stand idly on the river bank, gazing curiously at the approaching canoes, or to where, fording the stream, they seek the narrow pathway to their muddy stable yard.

We were now in a region where traces of man's handiwork appeared all about us. First there were log houses, then farms, with their network of fences on sloping hillsides, then churches, and finally villages. The log houses were rude affairs, mere shanties, small and confined, out of which peered many little faces. No extensive clearings seem to be attached to these homesteads, and the owners make their living by cutting out shingles back in the forests, and "hiring out" in the summer to the farmers down the river, and to the loggers in the winter. Their children seldom see the inside of a schoolhouse, and grow in profusion. Cases occur frequently on the St. John, we were told, of families which contain between twenty and thirty children.

A pretty sight on this river was that of a boy, about ten years old, dipping water out of the stream with a tin bucket into a large cask. The cask stood on a wooden sled, to which were attached two oxen, not more than two or three years old, and very small. They stood knee-deep in the stream, waiting patiently while their little driver, holding on to the cask with his left hand, bent his body over, dipped his bucket nearly full, and swung it easily up to the rim of the cask and poured its contents in. The dexterity and grace of his movements were quite marked, and he accomplished his task with all the ease of a full-grown man. As we came opposite, he paused, and let his arm and bucket hang by his side. He leaned against the cask and looked at us absorbed in apparent admiration, forgetting, or at least unaware, that he was quite as much an object of interest to us as we could be to him.

We spent that night under the hospitable roof of Martin Savage, one of the successful loggers and farmers of the St. John, opposite the mouth of the St. Francis River. Mr. Savage's hospitable and well-appointed house is always open to gentlemen tourists, to whom he is wont to extend a cordial welcome. In fact, the people of this section are, as a rule, all hospitable, and seem to enjoy quite well the visits of strangers from the outer world, and a chat with them about matters of general interest, of which, in their isolation, they hear but little.

The next day we passed Fort Kent, where stands an old block-house, a memento of the "Aroostook war," and before night were safely lodged in one of the hotels at Edmundston, in New Brunswick.

And here ended our journey by canoe. Since leaving Morris's, on the Penobscot, opposite the head of Moosehead Lake, we had come one hundred and sixty miles through the heart of the wilderness. The keen enjoyment of many hours had made ample amends for the few hardships we had undergone, while the lessons we had had of Nature's teaching will form a priceless treasure-book, of which, when we are far removed from her schoolhouse, we may turn the leaves anew, and read again and again the story we had conned. There we shall find an aromatic whiff from some distant forest, or the odor from the burning birch-log; we shall see ourselves at evening grouped about the campfire, listening to the tales of our Indian guide; the antlered moose, with lumbering gait, will rise before us, or we shall hear the midnight wailing of the loon. On these pages, too, we shall see the old and honored hills and mountains in their quiet dignity, unchanged from yesterday, unchanging tomorrow. In their sombre mantles, which shift to distant purple, they will remain unaltered, till stripped by man's insatiate greed. Even then, the noble outlines must remain. Above their topmost peaks summer clouds are floating, and on their forest slopes shadows rest awhile, only to follow others that have gone before. The ever-present and ever-changing waters at their base give out their glad welcome in rippling smiles, or in silence show the peace that lies in their tranquil bosom. The old familiar brook bubbles out its wonted song of yore, ever tumbling onward and disporting in wild glee among the rocks that strew its bed, or, nestling in some darksome pool, gives back the quiet image of the mossy bank. Nearby, the well-trod footpath meanders through the silent wood, whose scraggy arms enfold it, until, abruptly turning, in tangled thickets it is lost to view.

*

**2020 Edition**

**MAINE WOODLAND PATH**
One of the editor's favorite well-trod north woods trails.

*

# Appendix

§

## Appendix I. Indian Place-Names.

*

**IN** the autumn of 1881, with a view of collecting the unpublished Indian geographical names of that part of Maine represented on the accompanying map, and of learning from the Indians themselves the meanings of these names and of some others whose signification was not generally known, the writer visited the Indian island at Oldtown on the Penobscot. There he had an interview with John Pennowit, then eighty-seven years of age, the oldest hunter among the Penobscots, and acknowledged on all sides to be more thoroughly familiar with the Maine woods than any other member of his tribe. This interview was followed by a second in the next year, and by others with other Indians on different occasions, and the stock of names gathered in this way has been supplemented by short and desultory researches, on the writer's part, in the Abnaki Dictionary of Râle, republished by Dr. John Pickering, in the "First Reading Book in the Micmac Language," by the Rev. Mr. Rand, in the publications of Dr. J. Hammond Trumbull, and in some few other works that relate to the same subject.

The results of this research are here given, not because they are thought conclusive, but because of a wish to interest others in the same field, and to gratify a curiosity on the subject among the ever-increasing throng of visitors to the Maine forests. The names in the following list are in most cases given precisely as they were taken down from the mouths of the writer's informants, while in a few cases *b* and *p*, *d* and *t*, *gw* and *qu*, have been interchanged for the sake of preserving a more exact correspondence with the forms given by Râle. That this change is perfectly proper will doubtless be admitted, when one is told that the pronunciation of the Indians is very hard to apprehend, and that at times one cannot distinguish their *b* from *p*, their *d* from *t*, or their *m* from *n*. In fact, they probably do not perceive the difference themselves. Among the Penobscots, too, the a of Râle is often pronounced like *ĕ*, and *vice versa.*

Inasmuch as the writer knew absolutely nothing about Indian language at the time the greater number of these names were taken down, their value, if they have any, lies chiefly in the evidence they bear of the little change that has taken place in this language, or rather in this dialect of it, during a century and a half; a fact that, in view of the absence of any literature among this people, may well surprise us. While, too, the writer would not detract from the trustworthiness or accuracy of the philological explanations suggested, he must at the same time reserve the privilege of correcting any errors in them that further investigation may reveal to him.

The reader will soon perceive that a large majority of these names contain a component signifying either "lake," as *bégat, quasabam, gwasébem, gami-k, gamŏ-k, gamō-k, gŏmŏ-k, gōmä, gamä-k, gamaï-k, gamoi*-k, and *guamŏ-k*, (often preceded by a nasal sound), — "stream," as *tegwé-k, tegoo-k, ticook,* and *took,*— "mountain," as *wadjo* and *wodchu*, and the inseparable *adene,*— or "place," *ki-k, ke-k, koo-k, kŏ-k*, and *kea-g*, — most of which may take the diminutive *is, es,* or *sis.*

It will also be noticed that nearly all such names end with a *k* sound, as just written. This is the locative particle "at," which the writer has not translated in giving the explanations of the different place-names, because its force seems to consist merely in changing an indefinite general name to a particular one by localizing it, very much as we should distinguish between "a lake where there are eagles" and "Eagle Lake." The Indians never translate this particle.

The general explanations which follow the Indian names are to be understood, unless otherwise noted, as coming from Pennowit. Francis Nicholas, a noted hunter, and Steven Stanislaus, formerly Lieutenant-Governor of the Penobscots, are also at times quoted, and their initials follow the explanations derived from them respectively. The phraseology of the explanations is often that of the writer, read over to and confirmed by his respective informants. The Greek ω as here used is equivalent to *w* or *oo*; *ā* and *é* have the sound of *a* in "say," *ä* that of *a* in "what," *ē* that of *ea* in "seat," while small $^{n}$ after a vowel shows that the latter has a nasal sound. The character $^{c}$ is aspirate.

§

**Abacotné'tio** : a pond at the head of the north branch of the Kcttegwéwick, or West Branch of the Penobscot River.

**Abōcadnĕ'ticook** : the north branch of the Kcttegwéwick, or West Branch of the Penobscot, from its junction with the south branch; "stream between the mountains," or "stream *narrowed by* the mountains." Cf. *Nesowadnehunk*. We see here the in separable substantival *adene*, "mountain," and *tegωé-k* or *tegω-k*, (a specific) "stream." For the reason for applying this name to the stream from the *forks* up, see Kcttegwéwick.

**Abbahas, Abahos, Abahtacook** : a branch of the Madamiscontis; "a stream that runs parallel with a big river." (F.N.)

**Abōjēdge'wäk** : thoroughfare between North Twin and South Twin Lakes. "Two currents flow, one on each side of an island in the thorough fare." (S.S.) Cf. Eptchēdge'wäk.

**Āboljackarmé′gas** : a branch of the Penobscot, at the foot of Mount Ktaadn. It means "no trees, all smooth." From the almost uniform absence of the letter *r* in the names collected by the writer, it is possible that the r in this word should more properly be an aspirate, or *h*. Moreover, an intelligent Indian once told the writer that the proper form was *Aboljach'kamē'gek*, "bare, *or* bald place," in which *jach* was ejaculatory, and partook of the nature of an oath, as if one should say, "There is the *damned* bare place." In the latter form we see the component *kamighe*, "an enclosed place," with the locative *k*. Pennowit says there was a small point that was burnt over and left smooth.

**Āboljackarmégas'sic** : diminutive of the preceding. The mouth of this stream is just below that of the last named, and is a small, open, grassy meadow in the midst of the woods, pretty enough surely not to deserve "damning."

**Ahsēdäkwä'sic** : Turner Brook, on the upper St.John; "place on a stream where a stick or rod was pointing to some branch stream," i.e. as a sign for one to follow in that direction.

**Al'lagash** : the principal branch of the St. John; probably a contraction of *Allagaskwigam'ook* (q. v.).

**Allagaskwigam'ook** : Churchill Lake, on the Allagash; "bark-cabin lake." On a plan made in 1795 the name is spelled

*Lacassecomecook.* Another form is *Oolagweskwigamicook* (F.N.). Râle gives *pkωaha$^{n}$* for the bark used "à cabaner," and ωaraghéskw, "grosses écorces p'r brûler," but it is not at all unlikely that the latter is the spruce bark, which is often used for cabin-making. With the dialectic interchange of *l* for *r*, the etymology of the word becomes clear: *waraghéskω*, "bark," *wick*, "cabin," and *gami-k*, (a specific) "lake."

**Allah'twkikämō'$^{c}$ksis** : a pond near Soubungy Mountain; "ground where a good deal of game has been destroyed."

**Ambajee'jūs, = A$^{n}$bōjee'jūs** : a lake and falls on the Penobscot; so called, it is said, from two large round rocks in the lake, one on top of the other. Cf. *Abōjēdge'wäk.*

**Am'bajēmack'omas** : properly the name of Elbow Lake between North Twin Lake and the dam below it, but improperly applied to Gulliver Pitch, a fall on the Penobscot below Ripogenus Carry, where a man named Gulliver was drowned. The meaning is "little cross pond," which more fully appears, if we follow the analogy of one form of the name for Chamberlain Lake (q.v.), and write it *Ap'moojēnega'mis.* In these two instances and in *Parmacheene* (Lake), — which are the only three Indian names in Maine known to the writer in which *a'pmojeene* or *pemetsini* occur as components, — the force of the preposition appears to be that the general direction, or length, of the lake lies *across*, or *crosswise to*, the usual course or route of persons that go over it. Cf. *Numtsceenägä'näwis.*

**Androscoggin, Amoscommun** : a branch of the Kennebec; "somebody found something" (?).

**A$^{c}$pmoojēne'gamook**, or **Baamcheenun'gamook** : Chamberlain Lake, on the Allagash; "cross lake." It is from the words *a'pmoojēne* or *pemetsiniωi*, "crosswise," and *gami-k*, (a specific) "lake." Cf. *Am'bajēmack'omas* and *Numtsceenägä'näwis*.

**Aroo'stook, Aloostook, Oolastook** : "beautiful river" (Rand's Micmac Reader). Cf. *Wallastook.*

**Arumsunkhun'gan**. See *Nollommussocongan.*

**Aswaguscawa'dic** : a branch of the Mattawamkeag. *A*$^{n}$*zwazōguscawa'dik*, "a place where, on account of the distance, one drags his canoe through a stream, rather than carry it." Cf. *Usoogomŭsoogwĕdâmk*, "wading-across place, a ford" (Rand). Cf., also, *nedera*$^{n}$*sωgadω*, "je traine le canot dans le rapide" (Râle).

**At'tean** : a pond on Moose River; the name of an Indian family, and formerly of a chief.

**Ä'wangä'nis** : Priestley Lake, near the Allagash ; "lake or water reached from a river by cutting across country up a brook, and thence by land, instead of going around and up the outlet of the lake."

**Bas'kahé'gan** : a branch of the Mattawamkeag ; *Pas'kéhé'gan*, "a branch stream that turns right off down, while Piscataquis goes up straight" (F. N.). Pennowit says it is "where they made a weir to catch salmon, eels, anything."

**Beegwä'took** : Pushaw Pond, near Bangor ; "big bay place." *a*$^{n}$*bagωat*, "covert," or "cove."

**Bras'sua** : a lake on Moose River; said to mean "Frank."

**Caucomgo'moc, Kahkoguamook** : a lake near Chesuncook on the West Branch of the Penobscot; "big-gull lake." The word comes from *kaa'kω* (Râle), "the big, white gull," and *gaωi-k*, (a specific) "lake."

**Chemquasabam'ticook** : stream and lake tributary to the Allagash; "stream of a large lake," from *che*, "great," or "large," *pegωâsebem*, "lac" (Râle), and tegω-k, (a specific) "stream."

**Chesun'cook** : a lake on the West Branch of the Penobscot; "the biggest lake." The name may designate or be applied to the lake at its outlet, that being the point from which the Indian coming up the Penobscot would first see it. Its composition would then be from *che*, "great;" the root *sa*$^{n}$*k*, seen in *sa*$^{n}$*ghede*$^{c}$*tegωé*, "l'embouchure, sortie [de la rivière]," and in *sa*$^{n}$*ktâïïωi*, "où il finit d'être, rivière v. ruisseau, &c," i.e. its mouth, or in *nesa*$^{n}$*ki*$^{c}$*re*, "la terre sur le bord du fleuve" (Râle); and *ki-k*, (a specific) "place,"— "great-discharge place."

**Chicum'skook** : Grindstone Falls on the East Branch of the, Penobscot. The word is said to mean "big falls," and to be a mixture of two dialects, Maliseet and Penobscot (S. S.). It may come from *chi*, "big," *o*$^{n}$*bsk* or *ompsk*, "rock" or "bowlder," and *ki-k*, (a specific) "place," — "big-bowlder place."

**Chimkazaook'took** : one of the upper branches of the St. John; from *chi*, "great," *mkazéwighen*, (il est) "noir" (Râle), and *tegω-k*, (a specific) "stream," — "big black stream."

**Cob'scook** : a stream that empties into Passamaquoddy Bay; "falls, or rough water" (Passamaquoddy Indian). Cf. *Chicumskook*.

**Cussabex'is** : a pond near Chesuncook Lake; "where there is a big lake connected with a pond, so you can go up without poling" (F. N.). This explanation is evidently wrong, as it describes Moose Pond, below Cussabexis. May not the word be the name of the brook which flows out of Cussabexis, and be equivalent to *kesibecksis*, from kesitsωa$^{n}$n, "elle est rapide" (Râle), *nebi'* or *nebpĕ'*, "water," k, locative, and *sis*, the diminutive, — "the little swift water"?

**Ebee'me, or Ebee'min** : applied to a mountain, and to a gorge known as the "Gauntlet," north of Brownville, Piscataquis Co.; "where they get high-bush cranberries" (S. S.). Under "Les fruits des arbres," Râle gives *atebimin*, "gros co'e [comme] de gros poix, rouges," and *ibimin*, "rouges, mauvais."

**Éilandam'ookganop'skitschwäk** : "stair falls," On the East Branch of the Penobscot (S. S.).

**Eptchēdge'wäk** : thoroughfare between North Twin and South Twin Lakes; "where two currents coming from opposite or different directions meet." See *Abōjēdge'wäk*.

**Eskutas'sis** : a stream in Lowell; "small trout," from *skω'tam*, "truitte" (Râle), and *sis*, diminutive. Another form is *Skutarza*.

**Eskwes'kwéwad'jo** : Bald Mountain, north of Holden, Somerset Co.; "she-bear mountain." *Wadjo* is "mountain," while for "she-bear" Râle gives *atseskω*. Cf. Noosĕskω (Rand's Micmac Reader, page 44).

**Étasiï'ti** : Wilson Pond, near Moosehead Lake; "where they had a great fight," or "destruction ground."

**Hock'amock** : see *Nāmŏk'anŏk*.

**Katepskōnēgan** : falls and dead-water on the West Branch of the Penobscot ; "a carry over a ledge," from *kat, keht*, or *k't*, "big," *peskω*,"rock" or "ledge," *ωni'gan*, "carry,"— "big-ledge carry."

**Kawäps'kitchwäk** : MachiasWest River; "sharp rough rips" (F. N.); "rocky stream" (Passam. Indian).

**Ken'nebec** : "long river"; from *kωné*, "long," *nebpĕ'* , "water," and *ki*, "land,"or "place,"— "long-water place."

**Ket'tegwé'wick** : "the West Branch of the Penobscot"; formed from *ket* or *keht*, "great," *tegωé*, "stream," and wick, "place." This name and that of the East Branch, Wassā'tegwé'wick (q. v.), are applicable to the respective streams at their junction. The Indian coming up the river, when he has arrived at Nicketow, "the great forks," decides which branch he will follow, whether the western or "main branch," *Kettegwé*, or the eastern, "where he will find fishing [by torch-light]," *Wassā'tegwé*.

**Kin'eo** : a mountain in Moosehead Lake; said by a St. Francis Indian to mean "high bluff."

**Kla$^{n}$ganis'secook** : first falls on the Mattawamkeag River above the village of that name; "narrow like a door" (F. N.). Cf. kla$^{n}$gan,"*porte*"(Râle).

**Kōkadjeweemgwasebem** : Roach Pond, near Moosehead Lake; from *kok*, "kettle," *wadjo*, "mountain," and *pegωâsebem*, "lake," — "kettle-mountain lake."

**Kokadjeweemgwa′sebem'sis** : Spencer Pond, near Moosehead Lake; diminutive of the foregoing, "kettle-mountain pond."

**Kōkad'jo** : the more westerly of the Spencer Mountains, near Moosehead Lake ; "kettle mountain."

**Ktaadn** : in Piscataquis Co.; "the biggest mountain"; from *ket* or *k't*, "big," and the inseparable *adene*, "mountain." In Gyles's "Captivity" it is called "Teddon." Cf. Uktŭtŭnook, "the highest mountain" (Rand).

**Kwānä'taco$^{n}$gōmah'so** : Poland Pond, a tributary to Caucomgomoc Lake.

**Kwānōk'sa$^{n}$gäma'ĭk** : Loon Lake, near Caucomgomoc; "peaked pond," i. e. pointed at each end.

**Kwānō′sa$^{n}$gama'ĭk** : Webster Lake, on the East Branch of the Penobscot. The meaning is said to be the same as in the preceding. *Quœre*, whether *k* should be inserted here after *o*, or dropped from the other word, or whether there is a slight shade of difference in the significations of the two words ?

**Kweueuktōnoonk'hégan'** : Moose River; "snow-shoe river"; so called from that part of it above Attean Pond, where it bends like the frame of a snow-shoe (St. Francis Indian).

**Lapompique,** or **Lapompeag** : a branch of the Aroostook; "rope-stream, i.e. crooked" (P. N.).

**Lunksoos** : a mountain and stream on the East Branch of the Penobscot; "Indian devil," or "catamount."

**Macwa'hoc**: corrupted from Macwäkook, or Mackwokhok; "bog brook" (F. N.). According to Pennowit, however, it should be Temahkwi'cook, "beaver place"; from *tema$^{c}$kωé*, "beaver," and *ki-k*, (a specific) "place."

**Madawas'ka** : a branch of the St. John, and a branch of the Aroostook. *Ma$^{n}$dawas'kĕk*, "porcupine place" (Maliseet Indian). Râle gives *Ma$^{n}$daωessω* for "porcupine." Cf. *Madawescac*; *Mădawiskâk*, "where one river enters another" (Rand).

**Maddun'keunk, Medun'keeunk** : see *Namadunkeehunk.*

**Mahklicongo'moc** : Pleasant Lake, near the Allagash; "hard wood-land lake." Cf. Râle's *ma'ri$^{c}$kωk*, "lieu où il n'y a q du bois franc, seu, où il n'y a point de sapinage;" also, kwĕsow- mălegĕk, "a hard-wood point" (Rand).

**Mahkōnlah'gok** : the "Gulf" near Katahdin Iron Works; "a hole in the river" (S. S.).

**Mahnagwä'nēgwa'sebem** : "Rainbow Lake," near the Penobscot, opposite Mount Ktaadn. It is more than likely that this is a mere translation of the English name of the lake, and therefore not a genuine Indian place-name.

**Mahnēkébahn'tik** : Caribou Lake (near Chesuncook), or a place near its outlet, "where they used to get cedar bark for packs, &c. in going down the West Branch." Or, "where they got 'wycobee,' or leather-wood" (S. S.), the *wighebimisi,* "bois blanc," of Râle.

**Matagoo'dus** : a tributary of the Penobscot ; "bad landing" ; "bad landing for canoes " (F. N.) ; *Mategwé-oo-dis*, "meadow-ground" (S. S.). Cf. *agwiden*, "canot" (Râle), and *matta* or *matsi*, "bad."

**Matamiscon'tis** : a branch of the lower Penobscot. Dr. J. Hammond Trumbull thinks this word represents "*met-a*$^{n}$*msωakka*$^{n}$*tti*, 'a place where there *has been* (but is not now) plenty of alewives,' or to which they no longer resort." Coll. Conn. Hist. Society, Vol. II. p. 25.

**Matanau'cook**, *Matana'*$^{n}$*cook* : a branch of the lower Penobscot; "place of bad islands." Its composition is probably from *matsi* or *matta*, "bad," *mena'han* or *menan*, "island," and *ki-k*, (a specific) "place."

**Mäta$^{n}$ga'mook** : Grand Lake, on the East Branch of the Penobscot; "old, second-class lake." Pennowit explained by saying that hawks used to breed there in great numbers, and they killed off the ducks and partridges so completely that when the hunters came along they could not find food enough for their subsistence; that the ledges of the mountain "Horse's Rump" were covered with the feathers of the game thus destroyed, and the Indians, with contempt perhaps, called the place "old." The word then probably comes from the root of *méta*$^{n}$*dam*, "il est vieux, il ne peut plus aller nul part" (Râle), and *gami-k*, (a specific) "lake,"— "the old, exhausted lake."

An Indian, and an intelligent one too, on being asked by the writer to explain the meaning of this word, said, after some hesitation, that he could not do so in English, but afterwards on being pressed to do the best he could, said it meant "dirty, dusty,—

old," conveying the idea of something that is laid aside as unfit for use, and on which, therefore, the dust soon collects.

**Mata$^n$ga'mooksis** : Second Lake, just above Grand Lake. The word is a diminutive of the foregoing.

**Matawam'keag or Mattawam'keag** : a branch of the Penobscot; *Ma$^n$dä'wamkēk*, "down where a stream empties into the main river," and forms a pointed gravel or sand bar below its mouth, *connected with the main land.* A St. Francis Indian once explained to the writer the word mata$^n$bah', as "we are over into a better place " (cf. Râle, matta$^n$be, "il va au bord de l'eau," and *meta$^n$béniganik*, "au bout de delà du portage"), and *mata$^n$wam*, as "we are over the sand (or gravel) bar." Hence *Matawamkeag* should seem to mean "place beyond the sand (or gravel) bar," and should seem to come from the root of *mata$^n$be*, "beyond the end of," *am* or *um*, "sand " or "gravel," and *ki-k*, (a specific) "place." Cf. *Pamedomcook*.

**Meduxnekeag**, *Mēduxnee'kek* : a tributary of the lower St. John ; "where the people go out," i.e. from the interior or woods.

**Megkwah'lagas** : locality probably on or near the lower Penobscot; "red hole (on an island)." Râle gives *teta$^n$ωaragat*, "trou dans le bois, à la cabane ,&c," and the root *mkω* or mω$^c$kω, "red." Cf. *Mēgwasaak*, "red rock" (Rand).

**Megkwäk'ä$^n$ga'mik** : Mud Pond, at the head of the Allagash; "marsh pond" ; from *megωak*, "marécage, de l'eau des terres [?]" (Râle), and *gami-k,* (a specific) "lake."

**Megkwä'kä$^n$ga'mocsis** : Harrow Lake, near the Allagash; diminutive form of the preceding word.

**Mēla$^n$pswa$^n$gä'moc** : corrupted to *Meloxswangarmo* : Joe Merry Lake, near the Penobscot; "large-rock lake" (F. N.). A$^n$ps, from *peskω*, means "rock" or "ledge, "while *mēla$^n$ps* appears to mean "rocks of various shapes." Cf. *Milapskegĕchk'*, "abounding in rocks of all shapes and sizes" (Rand).

**Menhanee'kek** : Ragged Lake, near Chesuncook ; "place of many islands;" from *mena'hanωk*, "islands" (Râle),and *ki-k*, (a specific) "place."

**Mésak'kétésa'gewick** : the Socatean, a stream tributary to Moosehead Lake. The shorter word is probably a contraction of the other, being the second, third and fourth syllables of it, *sakkété*. According to Pennowit it means "half burnt land and half standing timber with the stream separating them." Cf. *nesakké*, "je suis debout" (Râle):— *wick*, "place."

**Meskaskeeseehunk** : north branch of the Mattawamkeag River; "little spruce brook" (F.N.). Cf. *messkask*, "pin rouge" (Râle).

**Meskee'kwägämä'sic** : Black Pond on Caucomgomoc stream; "grassy pond" ; from *meski'kωar*, "herbes" (Râle), *gami*, "lake," *es*, diminutive, and *ik*, locative.

**Mgwasebem'sistook** : Russell Stream, north of Moosehead Lake; "stream of a little lake"; from *pegωâsebem*, "lac" (Râle), *sis*, diminutive, and *tegω-k*, (a specific) "stream."

**Millinŏ'kett, Milno'kett** : lakes on the Penobscot and Aroostook respectively; by Pennowit pronounced as if written *Millnah'gkek*. It is said to be the equivalent of the Maliseet *millŏg'kami(k)*, which is a lake that has many irregularities in the way of points, coves, ledges, and islands. "If you ask me what kind of a lake Moosehead Lake was, I say 'millŏg'kami,' — i.e. it has no shape" (Maliseet Indian). Cf. *Milpāāchk*, "having many coves" (Rand).

**Mi'seree** : a pond and stream that empty into Brassua Lake; often spelled on old maps "Misery," but by some persons thought to be Indian. Râle gives *mesairi*, "bien," and *mesairedωr*, "plusieurs choses," from which may have come that part of the word that still remains.

**Misspeck'y, Moosepeck'ick** : part of the coast west of Machias. It probably means "overflowed" ; from *nemissbeghesi,* "je suis mouillé" (Râle). Cf. Mĕspaak, "overflowed (by the tide)" (Rand).

**Mkazaook'took** : Little Black River, a branch of the St. John, and Pine Stream, a tributary of the West Branch of the Penobscot; from *mkazéωighen*, (il est) "noir," (Râle), and *tegω-k*, (a specific) "stream," — "black stream."

**Molun'kus**, *Mōlun'kes* : a branch of the Mattawamkeag; "a short stretch of high land on a small stream." Also, "high bank on each side of the stream" (F. N.).

**Mooseleuk** : a branch of the Aroostook; "moose place." For "moose" Râle gives *mωs*.

**Moskwaswa$^{n}$ga'moc** : Shallow Lake, near Caucomgomoc; "muskrat lake"; from *mωskωéssω*, "rat musqué" (Râle), and *gami-k*, (a specific) "lake."

**Moskwaswa$^{n}$ga'mocsis** : Dagget Pond, on the stream just below Shallow Lake ; diminutive of the preceding.

**Mskwamāgwēsee'boo** : Hale Brook, on the upper Penobscot ("South Branch") ; "salmon brook" (Penobscot Indian); from *meskωamegωak*, "saumons," and *sipω*, "rivière" (Râle).

**Munolammonun'gun** : the west branch of Pleasant River, Piscataquis Co.; "very fine paint, or place where it is found, or great quantity of it." (See letter of Moses Greenleaf in a pamphlet entitled "The First Annual Report of the American Society for Promoting the Civilization and General Improvement of the Indian Tribes in the United States." New Haven, 1824.) The name takes its signification from the iron ore found in Katahdin Iron Works township, near the stream, this ore being sometimes of a bright vermilion color, and used, when ground up, for paint. Cf. *Oolammono*$^n$*lgamook.*

**Munsun'gan** : lakes at the head of the Aroostook. Pcnnowit at one time said *mu*$^n$*asu*$^n$ meant a "cut," and that *munsungan* meant "where they killed a good many moose and cut off streaks of fat (somewhere) between the shoulders." Cf. *Mŭnow*, "the fat of a bird" (Rand). At another time Pcnnowit explained the word as "where they spear (any kind of) fish."—*Ma*$^n$*sungun*, "where they speared salmon." (F.N.)

**Musquacook** : a tributary of the Allagash ; "birch bark place"; from *maskωé*, "ecorce de bouleau à cabaner" (Râle), and *ki-k*, (a specific) "place."

**Nahmajim'skicongo'moc** : Haymock Lake, near the Allagash; "lake of the dead-water that extends up into the high land."

**Nahmajimskit'egwek** : Smith Brook, Eagle Lake on the Allagash; "the dead-water extends up into the high land."

**Nahmakan'ta** : a lake and its outlet, tributary to the Penobscot; probably from *namés*, "fish," and *ka*$^n$*tti*, "there are plenty," — "where there are plenty of fish." The word *namés* is explained by the hunters among the Penobscots as "lakers" or togue, the *largest* of their lake fish, although Dr. Trumbull inclines to the belief that, as the form of the word is a diminutive, the fish designated by it must be of the "smaller sort."

**Nala'seema**$^n$**gamōk'sis** : Shad Pond, on the Penobscot ; "resting-place (after poling up the river)." Cf. *Nollesemic.*

**Nallagwā'gwis, Narraguay'gus** : river and bay .on the south eastern coast of Maine : "something breaks that you cannot fix."

**Namadun'keeunk** : "the stream is level and suddenly one comes to a swift place up which one must pole." — "It goes up rapid from the mouth of the brook." (F.N.) — "The brook runs up to the 'horseback.'" (S. S.) — Cf. "*Nĕmtâkâyak'*, 'it extends straight up rising ground' (you looking *up stream*, of course, in all such cases, and there being a long reach of rapids)." (Rand.)

**Nāmŏk'anŏk** : an island in the Penobscot above Oldtown, near Mohawk Rips; "high land, — kind of a lump." (F. N.)

**Nelhudus** : see *Nulhedus*.

**Nesowadnehunk'** : stream near Mt. Ktaadn; "the mountains from Ktaadn, *that* stream runs among them." This word is probably from *ntsaωiωi*, "au milieu" (Râle), or the more modern form *nesowāwi*, the inseparable *adene*, "mountain," and *hunk*, "brook." Just what is the difference in meaning between this word and *Abocadneticook* is not clear, unless the former signifies that the stream and its branches wind indefinitely among a group of mountains, and that the latter, at some part of its course, flows between mountains, or is hemmed in by them, as is actually the case with these streams, respectively. The explanation given under *Abocadneticook* — "narrowed by" — may have been meant for "hemmed in," or may have referred to the river *valley* more than to the stream itself.

**Nesun'tabunt** : a mountain near Nahmakanta Lake: "three-headed"; from *nass*, "three," and *a$^{n}$tep*, "head" (Râle).

**Nik'etow, Nicketou** : the junction of the East and West Branches of the Penobscot; "the forks." Cf. *Nĭktaak*, "river- forks" (Rand), and *niketaω'tegωe'*, "rivière qui fourche" (Râle).

**Nik$^{c}$a$^{n}$a'gamäk** : another name for Ragged Lake, near Chesuncook; "upper lake." Cf. *nikka$^{n}$ni$^{c}$ra*, "il marche le 1$^{r}$ par eau" (Râle), the root of which is evidently here joined with *gami-k*, (a specific) " lake."

**Nohlka'ïmana'han** : Deer Island, Moosehead Lake ; from *nωrkaω* (pigé), ("plat côté de) chevreuil" (Râle), with the interchange of *r* for *l*, and *mana'han*, "island." It is more than probable that the Indian name is merely the translation of the English, and not original.

**Nōlanga'moik** : Ripogenus Lake, on the Penobscot ; "resting-place (after the long carry below it)."

**Nolle'semic,** ***Nōla'sēmik*** : a lake near the Penobscot; similar in meaning to the preceding. The name seems to have been wrongly transferred to this pond from Shad Pond, Nalaseema$^{n}$gamōksis, for we find the latter on old plans designated "Noleseemack" and "Nollesemeck." See Land Office Records, B. 2, PI. 25, and B. 6, Pl. 1.

**Nollommussocon'gan** : an island just below the mouth of the Stillwater, Penobscot River; "where they catch alewives" (F.N.). Cf. *a$^{n}$msω* (Râle). Pennowit gave the form *Nolu$^{m}$sōkhun'gun*, for the same island, but did not explain it further than to say that the Indians used to stop there to "hunt."

**Nuka$^{n}$conga'moc** : Clear Pond, at the head of Musquacook Stream, Piscataquis Co.; "upper or head-water pond." Cf. *Nik'a$^{n}$agamäk*.

**Nulhe'dus** : a branch of the Penobscot, near Moosehead Lake. On a map made in 1815, the name is spelled "Nalla Hoodus" (Land Office Records, B. 5, Pl. 7). The form *Nallahoot'da* was given the writer by a St. Francis Indian, who said the stream was so called from the peculiar outline made by it with the river at and above their junction, the outline of a loop. *Nallahoodus*, "a fall on each side." (F. N.)

**Numtsceenaga'nawis** : Elbow Lake ; "a little cross-pond" at the lower end of North Twin Lake. This form is like the second form of name for Chamberlain Lake, *Baamcheenun'gamook*, the first syllable being very hard to distinguish, when spoken by the Indians. Cf. *Nĕmchenokpāăchk*, "crosswise-lying lake." (Rand.) Cf. also *Am'bajēmac'komas*.

**Numdemō'ciss** : a stream in Washington Co. ; "where the suckers go up to spawn." (F. N.)

**Oloostook** : see *Wallastook*.

**Oolammono'$^{n}$gamook** : form given by Pennowit, and designates probably the pond near the furnace in Katahdin Iron Works, recently rechristened "Silver Lake." Under "Vermilion, pinture," Râle gives *ωra'ma$^{n}$*, which with *gami-k* would make "vermilion-paint lake." Cf. *Munolammonun'gun*.

**O$^{n}$zwazōge'hsuck** : Penobscot Brook, at the head of the West Branch of the Penobscot (" South Branch") ; "when they carry by there, they have to wade across 'quartering.'" (Penobscot Indian.) Cf. *Aswaguscawa'dic*. The idea of wading and dragging a canoe prevails in both explanations.

**P'ahn'moïwad'jo** : Squaw Mountain, Moosehead Lake; phâinem, "femme " (Râle), "what is of, or belongs to, woman," and ωadjo, "mountain." The Indian name is taken from the English, and, so far as the writer can learn, there is no distinctive original Indian name for the mountain.

**Pamedomcook, Pemidumcook** : a lake on the Penobscot; "bar or shallow place between two lakes." Also, "a gravel or sand bar runs into or through the middle of the lake." (S. S.) The idea seems to be that the bar has not necessarily any connection with the land. The word appears to come from pemaiωi, (which here has rather the force of *au travers de*, "through the midst of," or *à travers*, "crosswise," than of *de travers*, "aslant,") an Abnaki word for "sand," which seems to be *um* [cf. *pogomkek, pogumkek*, and *pemâmkek*, the last "a stretch of sand" (Rand)] and *ki-k*, (a specific) "place," — "the place where the sand stretches through or across [the lake]."

**Parmachee'nee** : a lake and stream of the Rangeley system. Cf. *A*[c]*pmoojēne'gamook*.

**Pasconga'moc** : Holeb Pond on the headwaters of Moose River. Synonymous with *Pescongamoc* (q. v.).

**Passadum'keag** : a branch of the lower Penobscot; "falls running over a gravel-bed." Also, "getting over the gravel-bars." (S. S.) Also, "at the head of the rips," or "one goes up an incline and comes to dead-water." (P. N.) Cf. Râle, *pa*[n]*sitsiωi*, "au de-là du rapide, de la chûte d'eau." Cf. also *Matawamkeag*.

**Passamaga'moc**, corrupted to **Passamagammet** : a lake and rapids on the Penobscot. See *Pescongamoc*.

**Patagum'kis** : a tributary of the Penobscot, "sandy round cove." Cf. *Pogomkĭgeâk*, "a dry sandy place." (Rand.) It may come from *petegω*, "round," *um*, "sand " or "gravel," *ki*, "place," and *es*, diminutive, — "little round sand-place," or "little place of round gravel."

**Patagus'sis** : Smith Brook, a branch of the Mattawamkeag. It seems also to be called *Mēsotoocus*.

**Pataquonga'mis** : Telosinis Lake, south of Chamberlain Lake; a lake on the lower Allagash, and a pond between Allagash and Chamberlain Lakes; "round pond" ; from *petegω,* "round," *gami*, "lake," and *es*, diminutive. According to Râle, however, *petegωi'ghen* means "it is round" (like a ball), while "round and flat" is a different word altogether.

**Pātāweekongo'moc** : the name by which the Indians designated Telos Lake, before the canal was made between it and Webster Lake ; "burnt-land lake."

**Pātā'weektook** : Ragmuff Stream, a tributary of the West Branch of the Penobscot; "burnt-land stream." Cf. *Pĕtāwagŭmegĕk*, "a charred grove," and *Pedawogunaak*, "a burnt-over place" (Rand).

**Penob'scot** : one of the principal rivers of Maine; pronounced by the Indians *Pa$^n$nauwmbs'kek*. According to Dr. Trumbull, "the first syllable, *pen* (Abn. pa$^n$na) represents a root meaning 'to fall from a height,' — as in *panntekω*, 'fall of a river' or 'rapids;'; pena$^n$ki, 'fall of land,' the descent or downward slope of a mountain," &c. Pa$^n$naumbsk is said to mean "a sloping rock, or one that is larger at the top than at the bottom." Again, *Pa$^n$naumbskek* is explained by a Maliseet Indian, long resident at Oldtown, as "there are ledges on each bank of a river, and just below them the river widens considerably." Cf. *Banooŏpskek*, "opening out among rocks" (Rand). *Pan$^n$naumbskek* refers to some *point on the river*, the stream itself being called by the Indians *Pa$^n$aumbskook'took*. Cf. *Pnapeskω* (Râle).

**Pes'conga'moc** : a small lake north of Pamedomcook, and very near the Penobscot. From this latter circumstance it takes its name, branch lake," from *peské*, "branch" (primarily "divided" or "split"), and *gami-k*, (a specific) "lake."

**Pes'kébé'gat** : Lobster Lake, near Moosehead ; "branch of a dead-water" ; from *peské*, "branch," and *bégat*, an inseparable for "water" (from *nebpe*, and literally "where there is water"). On this lake a long point of land juts out into the water and encloses on one side a large and deep cove, whence the translation "split or divided lake " may be a better one than that given by Pennowit.

**Pes'kebski'tegwek** : Soper Brook, Eagle Lake on the Allagash; "branch of a dead-water emptying into a lake." In this word appear *peské*, "branch," *ski*, or *skit*, a root applied to water in a stream, "at rest," and *tegω-k*, (a specific) "stream."

**Pes'kédō'pi$^{c}$kek** : Alder Brook, upper waters of the West Branch of the Penobscot; "branch of an alder-place" (Penobscot Indian). The form given by S. S. is *Pes'ke wydō$^{r}$pikek*, from *peské*, "branch," *ωdoppi*, "l'aune" (Râle), and *ki-k*, (a specific) "place."

**Petconga'moc** : pond at the head of the Allagash ; "crooked pond, or one that returns in the same direction in which it first ran." Cf. *Petkootkweâk*, "the river bends' round in a bow." (Rand.)

**Piscat'aquis** : a branch of the Penobscot; "little branch stream" ; from *peské*, "branch," *tegωe*, "stream," and *es*, diminutive.

**Pockwoc'kamus** : a lake or dead-water on the Penobscot, and Mud Pond southwest of Chamberlain Lake; "mud pond." We see here in composition *gami-es*, "little lake," i. e. "pond."

**Pōkŭm'keswa$^{n}$gamō'ksis** : Harrington Lake near Chesuncook; "a pond with a gravelly outlet." This word, although it designates the lake, is more properly applicable to a small pond at its outlet. The word is formed from *pōkum*, "dry sand," *ki-es*, "little place," *gami-k*, (a specific) "lake," and *sis*, diminutive, — "little-dry-sand (or gravel) place pond." Cf. *Pogomkĭgeak*, "a dry sand place." (Rand.)

**Pongokwä'hemook** : Eagle Lake, Allagash River; "woodpecker place." "When the woodpeckers first came from the west and rapped on the trees, the Indians heard them, and named the lake from them."

**Pōpōkŏm'ukwodchu'ssu** : Whetstone Falls, on the East Branch of the Penobscot. (S. S.)

**Pōtāwadjo** : near Pamedonicook Lake, on the Penobscot; "whale mountain." (F. N.)

**Pōtōbe$^{c}$k'** : Lily Bay, Moosehead Lake; probably nothing more than the generic name for bay. In it are distinguishable the root *pōtō*, "bulging," *nebpé*, "water," and *k*, locative, "where the water bulges."

**Psiscon'tic** : Brassua Lake, near Moosehead; "handiest place to build canoes." (?)

**Quā'kis**, or **Quakish** : a pond on the Penobscot above Nicketow. See *Nalaseema$^{n}$gamōksis*.

**Sabōta'wan** : the more easterly of the Spencer Mountains, near Moosehead Lake; "bundle or pack, — the end of it, where the strap is pulled together."

**Sä$^{n}$ghibpä'$^{n}$took** : falls between Chesuncook and Ripogenus Lakes ; "rough or hard falls" ; from the root of *sa$^{n}$'gheré*, "cela est dur," and *pa$^{n}$tekω*, "chute d'eau " (Râle).

**Sahbimski'tegwek** : Thoroughfare Brook, below Eagle Lake on the Allagash; "a branch or stream that empties between two large bodies of water."

**Sahkha'béhaluck'** : Moose River, Moosehead Lake; "more water flowing from it than from any other stream that empties into the lake."

**Sahkkahé'gan** : Telos Lake (ever since the cut was made); "water connecting with another body of water." See *Pātāweekongomoc*.

**Sapompeag** : should be *Lapompeag* (q. v.).

**Sāwada'bscook**, or **Sowadabscook** : a branch of the lower Penobscot; "place of large smooth rocks." Here we see *ompsk* or *auwmbsk*, "rock," and *ki-k*, (a specific) "place."

**Schoo'dic** : the name of several lakes and streams in Maine. Pcnnowit gives for it the form "Eskoo'tŭk, trout place."

**Sebam'ook** : Moosehead Lake. See following word.

**Sebec'** : a lake and stream tributary to the Piscataquis; "large body of water," or "extending water." The writer is very much inclined to the belief that this word comes from the root ωassā, "bright," *nebpĕ'*, "water," and *k*, locative; for we find in Râle, under "Clair," *ωasséghen*, "il est clair à travers ces arbres, il faut qu'il y ait là une rivière, lac, prairie, &c." A similar derivation would hold for *Sebay'gook* (or Sebago, ωassébégät), *Seba'mook*, and *Xsébem'*, the four forms being said by the Indians to have the same general meaning. A St. Francis Indian once told the writer that *Xsébem'* would be the exclamation he should use, if, on going through the woods, he should see the light grow bright through the trees, an indication that a pond was near.

**Seboo'is** : a lake and stream tributary to the East Branch of the Penobscot ; from *sipω*, "river," and *es*, diminutive, "little river."

**Seebä'ᶜticook** : Indian Pond, on the Kennebec, just below Moosehead Lake; "'logon' stream."

**Seeboo'mook** : Elm Stream, on the Penobscot, north of Moosehead Lake ; "my river" (F. N.). Each regular hunter and trapper has his own territory in the forests, on which it is considered a breach of "backwoods" etiquette for others to hunt or trap. This stream was probably within the district of some such hunter. *Sipω*, "river."

**Sisladob'sis** : a lake in Eastern Maine ; probably from *Si'galondo'pskes*, "rocky lake " (S. S.).

**Skit'ticook** : a branch of the Mattawamkeag ; "dead-water" ; from *ski*, or *skit*, "water" in a stream "at rest," and *tegω-k*, (a specific) "stream."

**Skowhē'gan** : a town and waterfall on the Kennebec ; "where the Indians used to wait for fish to run up, and to spear them as they went by." Cf. *ka"kskaωihigan* (Râle).

**Skutar'za** : see *Eskutas'sis.*

**Soca'tean** : see *Mésak'kétésa'gewick.*

**Sōghä'li manä'han** : Sugar Island in Moosehead Lake. ' This word is beyond doubt a translation of the English name, which was given to the island by Joseph Norris, surveyor, in 1827, on account of the large amount of sugar-maple on it.

**Tă<sup>c</sup>cook'** : a place about midway of the Indian island at Oldtown, on the west shore, at the rapids. The name was given to the writer by an intelligent Indian, who said it meant "waves." As a termination, it is generally written *ticook,* from *tegω-k,* and seems to refer in Maine place-names primarily to the ripples or waves made by rapid water.

**Temahkwé'cook** : see *Macwa'hoc.*

**Tlowā'wā'yĕ** : Third Lake, on the East Branch of the Penobscot.

**Tomhē'gan** : a stream which empties into Moosehead Lake; from *tōma'hégan',* "hatchet." The name is doubtless not of Indian application, or else it has lost part of its original form.

**Tulan'dic** : a branch of the upper St. John ; "where they make canoes" (F. N.). Greenleaf gives the alternative name "R. du Canot,"

**Umbazook'skus** : a tributary of the Penobscot, at Chesuncook Lake ; "meadow place."

**Umcoleus** : sec *Umcolquis.*

**Umcol'quis** : a lake and stream tributary to the Aroostook; from "*umcolquesook,* whistling duck" (F. N.).

**Umsas'kis** : a lake on the Allagash; *Ansaskek*, "having opposite points which run out to meet one another" (Maliseet Indian). Very graphically described in Greenleaf's list (see reference under *Munolammonun'gun*) as "tied together like sausages."

**Unsun'tabunt** : a name found on old maps for Rainbow Lake; "wet head" (F. N.). Quœre, Can the name be a corruption of *Nesuntabunt*?

**Wahkasek'hoc** : on the Mattawamkeag River; "where moosehide frames were left, after the hides had been cut out." (F. N.)

**Wal'lastook, Woolastook** : the St. John River ; "stream where you get smooth boughs." Authorities generally think this word means "fine," "good," or "beautiful river." Cf. Mr. Rand's "St. John River, *Oolastook*, 'beautiful river'" ; also, *ωrastegω*, "la rivière de St. Jean " (Râle).

**Walleniptéwee'kek** : South Twin Lake on the Penobscot ; "round coves surrounded by burnt land." We see here the same component, *ptéweek*, as in Pātā'weektook (q. v.). For the other component, cf. *Wŏlnâmkeâk*, "a sandy cove," and *Wŏlnŭmkeajechk*, "a small sandy cove " (Rand).

**Wassā'taquoik** : a tributary of the East Branch of the Penobscot. See following word.

**Wassā'tegwé'wick** : the East Branch of the Penobscot; "anybody spearing," or, more correctly, "place where they spear fish." The root *ωassā*, from ωasséghen, "il est clair," or ωasséghen, (il est) "blanc" (Râle), primarily means "white," "bright," or "clear," and there are some who think *Wassā'tegwé'wick* means "place of the bright or sparkling stream." The secondary meaning of ωaasā, as given by Râle, appears in *ωasssénemaïωi*, "au au flambeau, avec une lumière," and again under "Poisson," where we find *nωassa,* " j'en prens au flam beau," *n* being the affix for the personal pronoun of the first person.

This secondary meaning of ωassā is the one applied to the East Branch of the Penobscot, from Niketow up, by all of the hunters among the Penobscot Indians that the writer has questioned about it. This stream has been noted for its salmon, which the Indians as a people spear at night, by torch-light. The writer has no hesitation then in accepting before all others the translation from *ωassa, tegωe*, "stream," and *wick*, "place," — "fish-spearing-stream place."

The form *Wassātaquoik* seems to be practically the same as *Wassātegwéwick*, *taquoik* taking the place of *tegωé-k*. Pennowit made no distinction between them, except to say repeatedly, on two occasions a year apart, that the latter was the name of the main East Branch, while the former, which he pronounced *wassā'tăcook'*, was exclusively the name of the smaller stream tributary to it.

**Wassŭm'kédéwad'jo** : White Cap Mountain, near Katahdin Iron Works ; "white sand mountain," because from a distance the bare spots of detritus on its summit look like sand (S. S). *Wassum* is from *ωassé*, "white," "bright," or "shining," and *um*, "sand " or "gravel." *Wadjo* is " mountain."

**Wa'toolwa$^{n}$gam'ook** : St. John Pond, headwaters of the St. John River; "pond where you keep cattle, sheep, caribou, moose," &c. i. e. " good hunting ground."

**Woboos'took** : Baker Stream, headwaters of the St. John. Cf. *Wobooĕk*, "the water appears white" (Rand).

**Woolas'taquāguam'** : south branch of the St. John. The name of the lake from which this stream flows has probably been abbreviated and otherwise changed, and applied to the stream. See following word.

**Woolas'tookwa$^{n}$guam'ok** : Baker Lake, near the head of the St. John River ; "lake of the stream where you get smooth boughs." Sec preceding word.

**Wydō'piklock, Wytopidlot** : a branch of the Mattawamkeag; "the river is broad, and there are no trees on its banks except *alders* (*wydō'pi*)." Cf. *ωdoppi* (Râle).

**Xsébem'** : Moosehead Lake ; "extending water." See the word Sebec.

***2020***: *(And the editor believes it to be fitting that the Indian name list ends with* ***Xsebem'****)*

# Appendix II. Cross Index

Cross-index to Indian names in the preceding list.

Alder Brook, Pes'kédō'pi $^{c}$kek.

Baker Lake, Woolas'tookwa$^{n}$guam'ok.

Baker Stream, Woboos'took.

Bald Mt., Eskweskwéwadjo.

Black Pond, Meskee'kwägämä'sic.

Black River. See Great and Little Black Biver.

Caribou Lake, Mahnēkébahn'tik.

Chamberlain Lake, A$^{c}$pmoojēne'gamook,.

Churchill Lake, Allagaskwigam'ook.

Clear Pond, Nuka$^{n}$conga'moc.

Dagget Pond, Moskwaswa$^{n}$ga'mocsis.

Deer Island, Nohlka'ïmana'han.

Eagle Lake, Pongokwä'hemook.

East Branch Penobscot, Wassā'tégwé'wick.

Elbow Lake, Numtsceenaga'nawis.

Elm Stream, Seeboo'mook.

Gauntlet, Ebeeme.

Grand Lake, Mata$^{n}$gamook.

Grindstone Ealls, Chicumskook.

Gulf, Mahkōnlah'gok.

Great Black River, Chimkazaooktook.

Hale Brook, Mskwamāgwēsee'boo.

Harrington Lake, Pōkŭm'keswa$^{n}$gamō'ksis.

Harrow Lake, Megkwä'kä$^{n}$ga'mocsis.

Haymock Lake, Nahmajim'skicongo'moc.

Holeb Pond, Pascongamoe.

Indian Pond, Seebä′$^{c}$ticook.

Joe Merry Lake, Mēla$^{n}$pswa$^{n}$gä'moc.

Lily Bay, Pōtōbe$^{c}$k'.

Little Black River, Mkazaook'took.

Lobster Lake, Pes'kébé′gat.

Loon Lake, Kwānōk'sa$^{n}$gäma'ĭk.

Machias West River, Kawapskitchwak.

Moose River, Kweueuktōnoonk'hégan, and Sahkha'béhaluck'.

Moosehead Lake, Sebam'ook, and Xsébem'.

Mud Pond, Megkwäk'ä$^{n}$ga'mik, and Pockwoc'kamus.

Penobscot Brook, O$^{n}$zwazōge'hsuck.

Pine Stream, Mkazaook'took.

Pleasant Lake, Mahklicongo'moc.

Pleasant River, Munolammonun'gun.

Poland Pond, Kwānä'taco$^{n}$gōmah'so.

Priestley Lake, Ä'wangä'nis.

Pushaw Pond, Beegwä'took.

Ragged Lake, Menhanee'kek, and Nik$^{c}$a$^{n}$a′gamäk.

Ragmuff Stream, Pātā'weektook.

Rainbow Lake, Mahnagwä'nēgwa'sebem.

Ripogenus, Nōlanga'moik.

Roach Pond, Kōkadjeweemgwa'sebem.

Round Pond, Pataquongamis.

Russell Stream, Mgwasebem'sistook.

Second Lake, Mata$^{n}$ga'mooksis.

Shad Pond, Nala'seema$^{n}$gamōk'sis.

Shallow Lake, Moskwaswa$^{n}$ga'moc.

Silver Lake, Oolammono'$^{n}$gamook.

Smith Brook, Nahmajimskit'egwek, and Patagus'sis.

Soper Brook, Pes'kebski'tegwek.

Soubungy Pond, Allahtwkikamōksis.

South Twin Lake, Walleniptéwee'kek.

Spencer Mts., Kōkad'jo, and Sabōta'wan.

Spencer Pond, Kokadjeweemgwa′sebem'sis.

Spider Lake, Allagaskwigamooksis.

Squaw Mt., P'ahn'moïwad'jo.

Stair Falls, Élandam'ookganop'skitschwäk.

St. John Pond, Wa'toolwa$^{n}$gam'ook.

St. John River, Wal'lastook.

Sugar Island, Sōghä'limanä'han.

Telos Lake, Pātāweekongo'moc, and Sahkkahé'gan.

Telosinis, Pataquonga'mis.

Third Lake, Tlowā'wā'yĕ.

Thoroughfare Brook, Sahbimski'tegwek.

Turner Brook, Ahsēdäkwä'sio.

Webster Lake, Kwānō′sa$^{n}$gama'ĭk.

West Branch Penobscot, Ket'tegwé'wick.

Whetstone Falls, Pōpōkŏm'ukwodchu'ssu.

White Cap Mt., Wassŭm'kédéwad'jo.

Wilson Pond, Étasiï'ti.

# Appendix III. Shorage Of Moosehead Lake

See note in text, Chapter 1.

In 1827 Joseph Norris surveyed the shores of Moosehead Lake, except those of Day's Academy Grant, from the north line of Bingham's Kennebec Purchase northward and around the lake down the east side to the northwest corner of Saco Free Bridge Grant, now a part of Greenville. The total measurement of this distance was 26,154| rods, or less than 82 miles. If we allow 30 miles for the shorage of Day's Academy Grant, and 35 miles for that of the other part omitted in Norris's survey, we have for the total shorage of the lake 147 miles. The distance apart of Norris's contiguous stations varied from a few rods to upwards of a mile; his total number of stations was 385, about 68 rods apart on an average. Hence we should allow something for distance gained by deviation or curvature of the shores. Thirty-eight per cent, or 53 miles, would seem to be a large allowance, and that would bring the total up to only 200 miles.

Again, from the north line of Bingham's Kennebec Purchase to the mouth of Moose River, Norris made the shorage 2797 rods (with 39 stations), and from Moose River to Socatean River 3,492 ½ rods (with 94 stations), or together nearly 20 miles. The sinuosity of the shore between these points is fully equal to its average sinuosity, and the distance between them in a straight line is 9.8 miles, or half the distance by the shore. The average width of the lake is not more than 5 miles; its length, 36 miles. The distance around it, in direct lines, would then he

36 + 5 + 36 + 5 = 82 miles, and twice this amount, 164 miles, would be the apparent shorage, to which we may add 36 miles, or nearly twenty-two per cent, for error, and our total then becomes 200 miles,— figures considerably below the popular estimates. According to Norris, the shorage of Sugar Island is 12 miles 219 rods ; that of Farm Island, 5 miles 14 rods ; that of Moose Island, 3 miles 301 rods ; and that of Sandbar Island, 1 mile 140 rods. The distance around the Mount Kineo "tract," through a pond southeast of the mountain, is 6 miles 136 rods.

* See Commissioners' Survey, A, Land Office, Augusta.

# Appendix IV. Soundings in Moosehead Lake

The soundings referred to were taken on several different occasions, those off the east and northeast face of Mount Kineo having been made by George V. Leverett, Esq., and the writer, assisted by Edward Masterman as guide; the others, by the writer and his Indian guide. The accompanying chart shows approximately the locality of each sounding. Those between Farm Island, Socatean Point, and the eastern face of Mount Kineo were made on "Yellow Tuesday," in August, 1881. At that time the temperature of the air was 78° (Fahr.), and that of the water at the surface of the lake, 66°.5.

The following table shows the temperature of the water at different depths, viz.: —

| Depth. | Temp. | Depth. | Temp. |
|---|---|---|---|
| Feet. | ° | Feet. | ° |
| 85.5 | 51.5 | 189.0 | 44 – |
| 166.0 | 45.5 | 181.0 | 44.0 |
| 188.5 | 45.0 | 191.0 | 43.75 |
| 210.0 | 44.0 | 194.0 | 44.0 |
| 208.0 | 44.0 | 194.0 | 44.0 |
| 211.0 | 43.5 | 183.5 | 44 + |
| 212.0 | 44.0 | 182.0 | 44.0 |
| 181.6 | 44 + | 205.0 | 43.5 |
| 105.0 | 47.5 | 193.0 | 43.5 |
| 63.5 | 51.5 | 142.0 | 46.0 |
| 55.0 | 58.5 | 98.5 | 50.0 |
| 113.0 | 46.0 | 26.0 | 64.0 |
| 126.0 | 45.5 | 105.0 | 46.5 |
| 154.0 | 44– | 180.0 | 43.5 |

# MAP OF FARM ISLAND BASIN
# MOOSEHEAD LAKE.

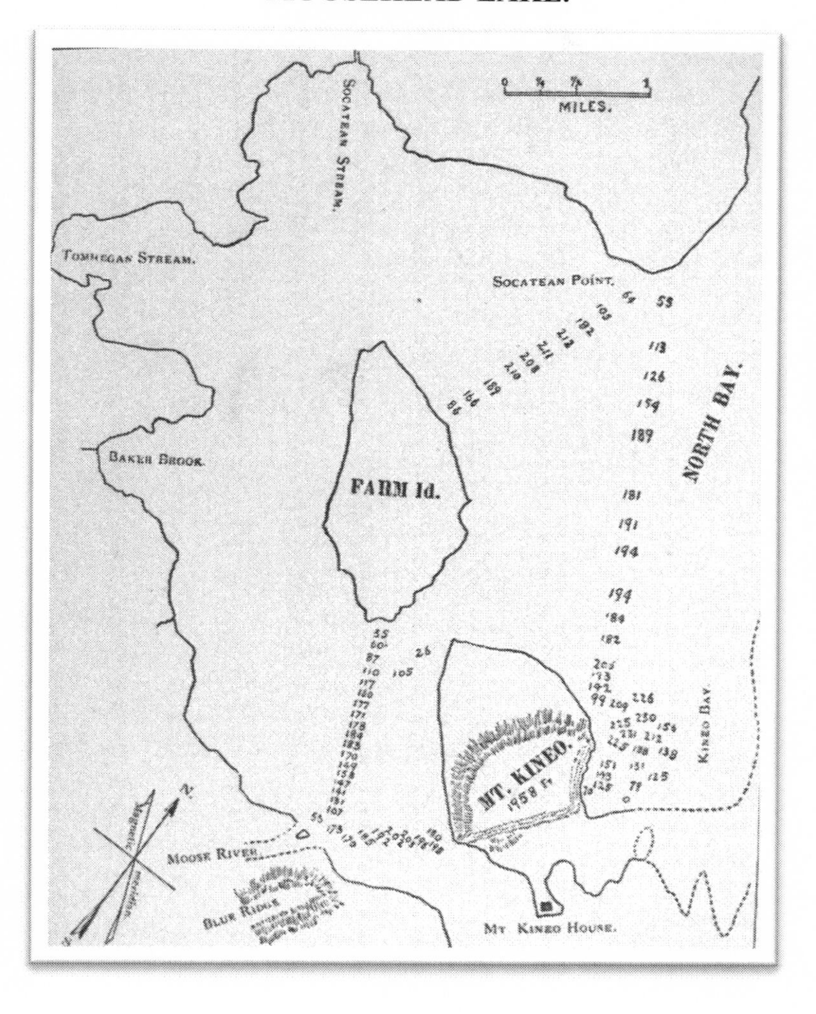

# Appendix V. Telos Canal

Until the year 1841 the waters of Chamberlain Lake flowed through a chain of lakes northeasterly into the Allagash and St. John Rivers. A large tract of fine timber country thus had no outlet to market except through the waters of New Brunswick. Just south of Chamberlain, and emptying into it, were two lakes, the upper of which, Telos, is only a mile from Webster Lake on the East Branch of the Penobscot. The latter lake is said to be forty-seven feet lower than Telos. A ravine, according to geologists the natural bed of a stream which in antediluvian times connected these two lakes, begins a few rods from the head of Telos and runs down into "Webster Lake. It was therefore perceived, that by an inconsiderable expenditure of money a thoroughfare for logs could be made between the two bodies of water, and that this would give the land-owners what they wanted, a direct route for their timber to a home market. Accordingly, in March, 1841, a dam was built at the upper end of Telos under the superintendence of Major H. Strickland, for Amos M. Roberts and another, who then owned Telos township. Some trees were taken out at that time from below the dam, and in the following April, or early in May, Chamberlain Dam was also built, under Major Strickland's orders. That spring the water ran over into Webster Lake and a successful "drive" was made through Webster Brook, for which purpose in part, or, according to Benjamin Dyer, for which purpose alone,[1] the Telos Dam had been built.

The next fall more trees were grubbed up, to a width of from forty to sixty feet, between the dam and Webster Lake, and a channel was dug between the dam and Telos Lake, ten or fifteen feet wide and thirty rods long. In the winter of 1842, more work was done below the dam, the digging being irregular, and more to guide the course of the water than for any other purpose, and that it might not spread over too wide a surface. The statement of Springer, in his "Forest Life and Forest Trees," page 204, that" originally the canal was three hundred rods long by four wide, and four feet deep," is misleading, and does not seem to be sustained by the evidence.

Telos township, and with it the cut and dam, were soon afterwards sold by Roberts to Rufus Dwinel, of Bangor. The advantages of the cut became apparent at once, and its possessor determined to reimburse himself for the outlay it had cost him. Certain persons who had cut logs on Chamberlain Lake were asked to pay from thirty-five to fifty cents per thousand feet for the privilege of driving them through the cut. The price demanded was deemed extortionate, and several of the loggers, Cooper & Co., Leadbetter of Bangor, Hunt of Oldtown, and others, refused to pay them, and, it is claimed, threatened to put their logs through by force. Whether any such threats were really made should seem to be doubtful. At any rate, the pamphlet heretofore quoted negatives that hypothesis. However, bound to protect his rights, Dwinel went "down river," got a hundred resolute men, armed them with knives, picks, hand spikes, and axes, and put them on guard at the cut. They blocked up the outlet with hemlock trees, and when the other loggers, with their "drivers," came along, they were surprised and chagrined to find themselves outnumbered and powerless. Finally, all agreed in writing to pay the required toll, and their portion of the expense of the hundred guards, except Leadbetter, who left his "drive" in Telos thoroughfare. This little episode is known as the "Telos War."

In 1847, after much opposition, a charter was obtained for the Telos Dam, and the toll now allowed for passage through it is twenty cents per thousand feet.

For much of the information contained in the foregoing, the writer is indebted to Calvin Dwinel, Esq., who is a brother of the late Rufus Dwinel, and who was personally on the scene, and participated in the events described.

1. See pamphlet entitled, "The Evidence before the Committee on Interior Waters, on Petition of William II. Smith, &e., &c., for leave to build a Sluice way from Lake Telos to Webster Pond."

# Appendix VI – Hubbard's Map

## Addition to the 2020 Annotated Edition

"Map of Northern Maine - specially adapted to the uses of sportsmen and lumbermen."

Available online from Map Collections

Map of Northern Maine 1900, Lucius L. Hubbard

No Copyright - United States.

A file of this map is available for download online with excellent ability to zoom-in on the details Hubbard painstakingly added to what was a poster-size print.

## Appendix VII - Maps of the Journey (2020 Edition)

The following cropped map sections are from Hubbard's map from Appendix VI. Some of the waterways and lands have been renamed since his survey. The explorer is advised to reference more recent maps for navigational purposes.

| Map | Location | Chapter | Description |
|---|---|---|---|
| 1 | ⓪ | 1 | Moosehead Lake, West Cove, and Town of Greenville |
| 2 | ① | II | Starting from Kineo |
| 2 | ② | II | Northeast Carry |
| 2 | ③ | II | Lobster Stream, Lobster Lake, Little Lobster Lake |
| 3 | ④ | III | On the Penobscot to Ragmuff Stream |
| 3 | ⑤ | III | Pine Stream, Pine Stream Falls, to Chesuncook Lake |
| 4 | ⑥ | IV | Up the Umbazookskus Stream, to Mud Pond Carry, to Chamberlain |
| 4 | ⑦ | V | To Chamberlain Lake |
| 5 | ⑧ | V | Dam between Chamberlain Lake and Eagle Lake. |
| 5 | ⑨ | VI | Smith Brook<br>Diversion to Haymock Lake |
| 5 | ⑩ | VI | Thoroughfare Brook to Churchill Lake |
| 5 | ⑪ | VII | Spider Lake (Allagaskwigamooksis) to Pleasant Lake. |
| 6 | ⑫ | VIII | To Harrow Lake (Megkwakagamocsis) |
| 6 | ⑫ | IX, X | Bog Brook |
| 6 | ⑬ ⑭ | XI | To 4th Lake (Musquacook Lakes)<br>Joe & Silas depart for food at Depot Farm ◊.<br>Through 4th to 1st Lake. |
| 7 | ⑮⑯ | XII | Musquacook Stream to Allagash River to Moirs |
| 8 | ⑰⑱ | XII | Moirs to St. John |
| 8 | ⑲ ⑳ | XII | St. Francis River to Edmundston |

## Map 1 – Moosehead Lake and Greenville.

## Map 2 – Mt. Kineo to Lobster Lake

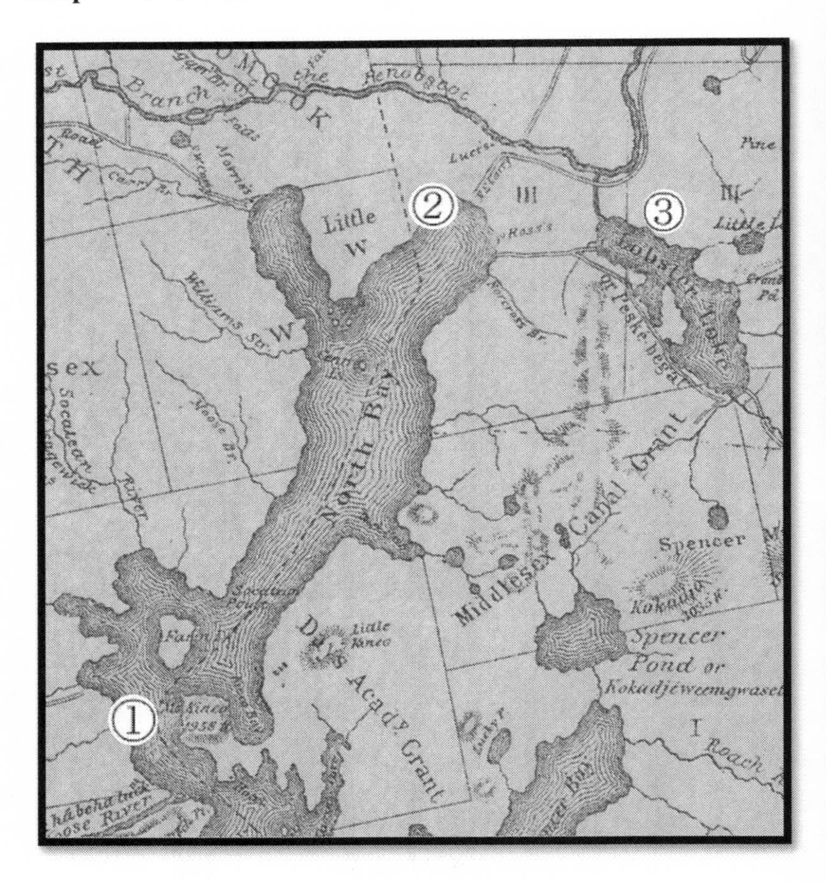

## Map 3 – On The Penobscot – Lobster Stream to Ragmuff to Pine Stream

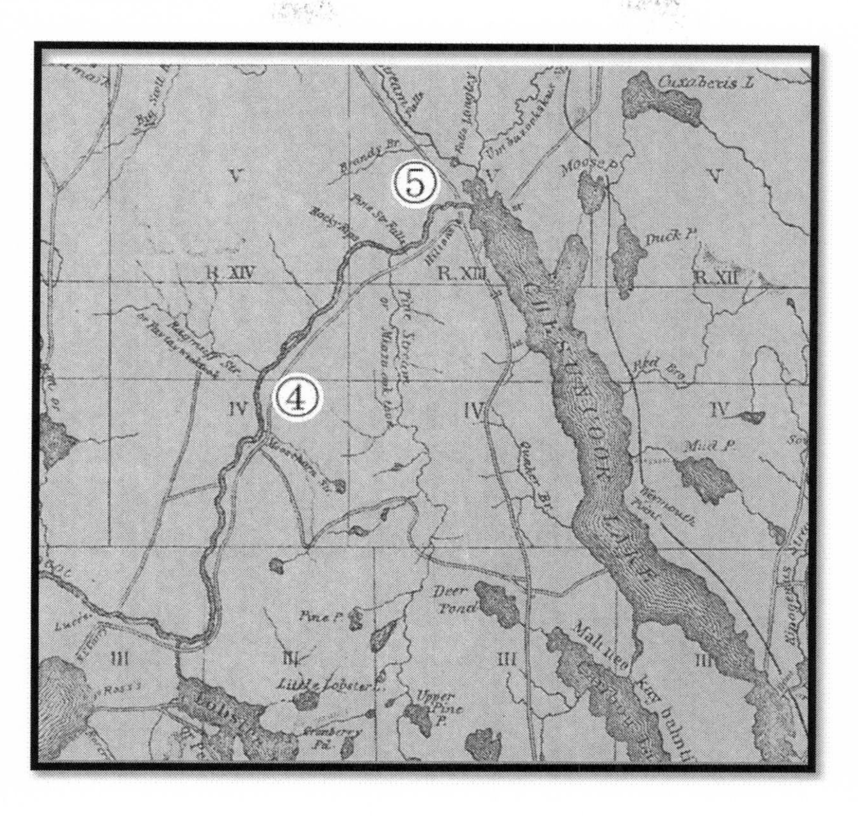

## Map 4 – Umbazookskus Stream, Mud Pond Carry, to Chamberlain Lake

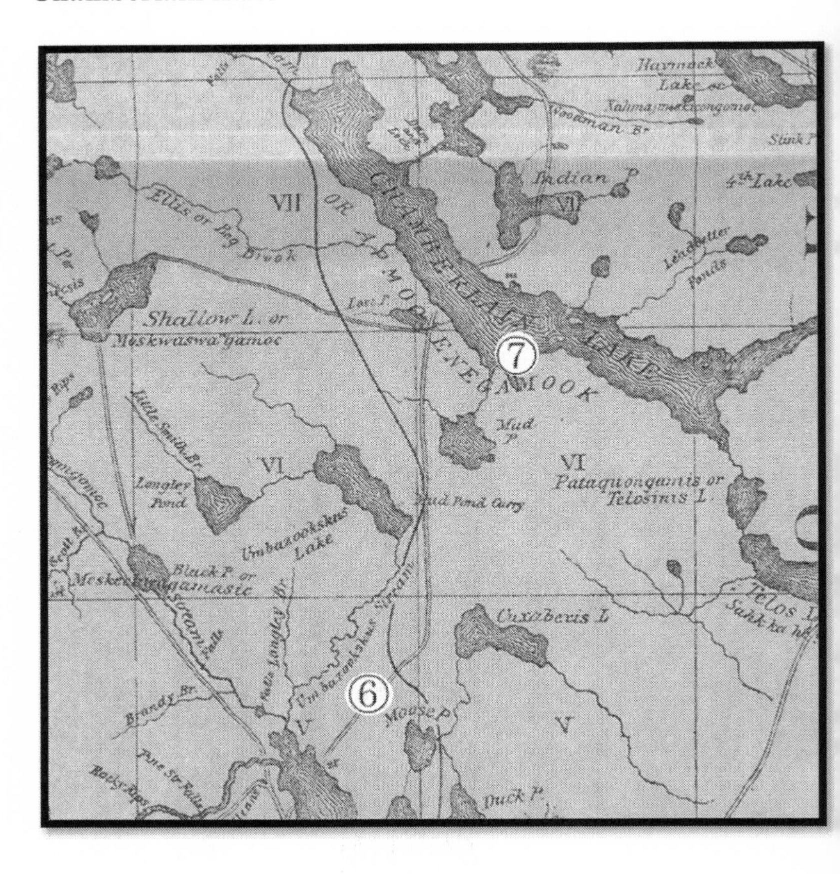

## Map 5 – Chamberlain Lake to Pleasant Lake

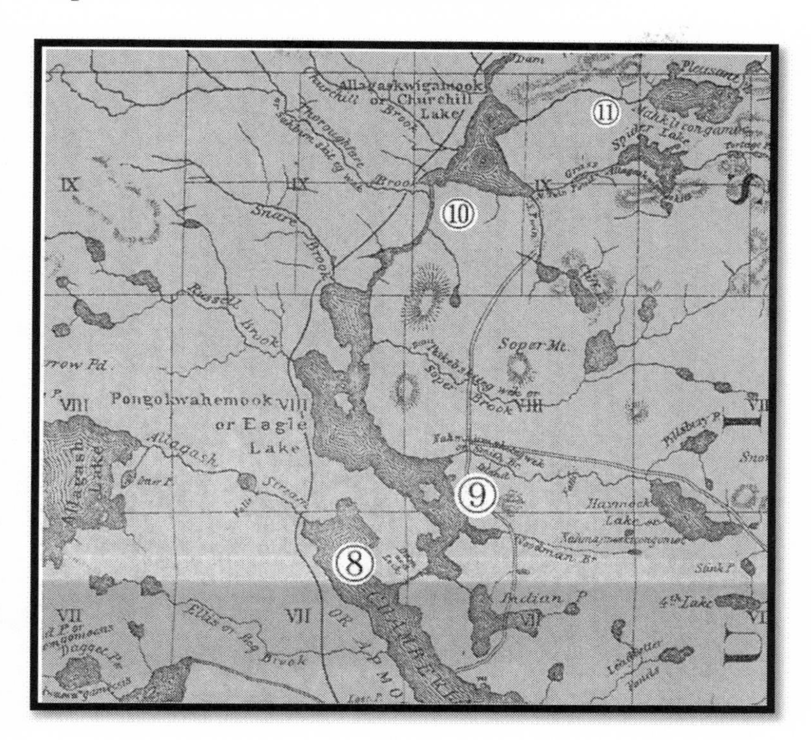

## Map 6 – Harrow Lake to 1st Lake

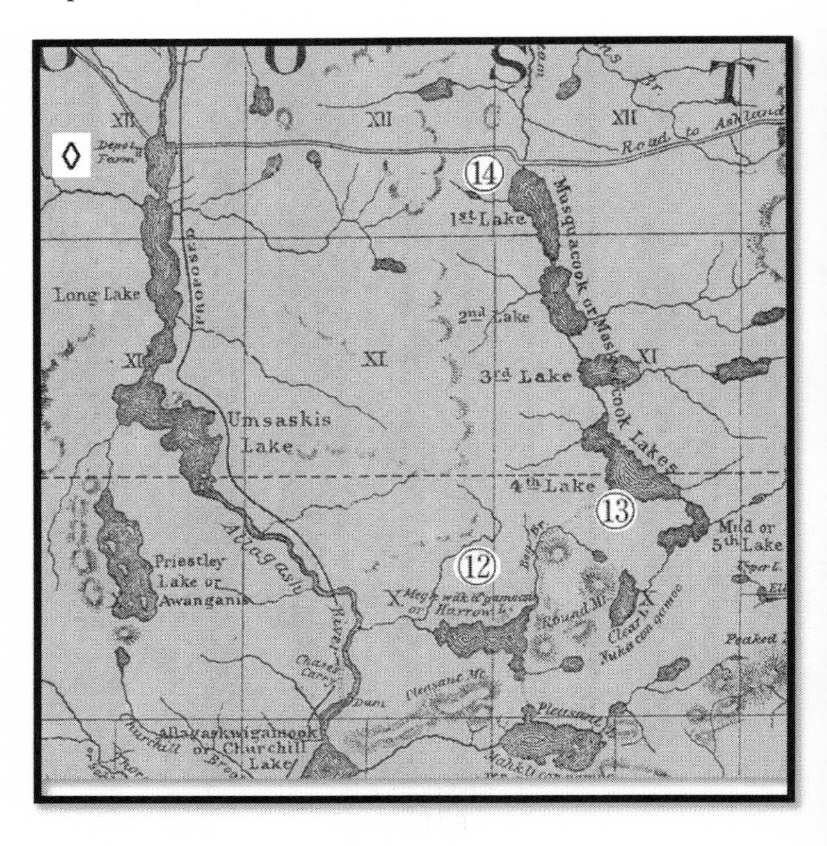

## Map 7 - Musquacook Stream to Moirs

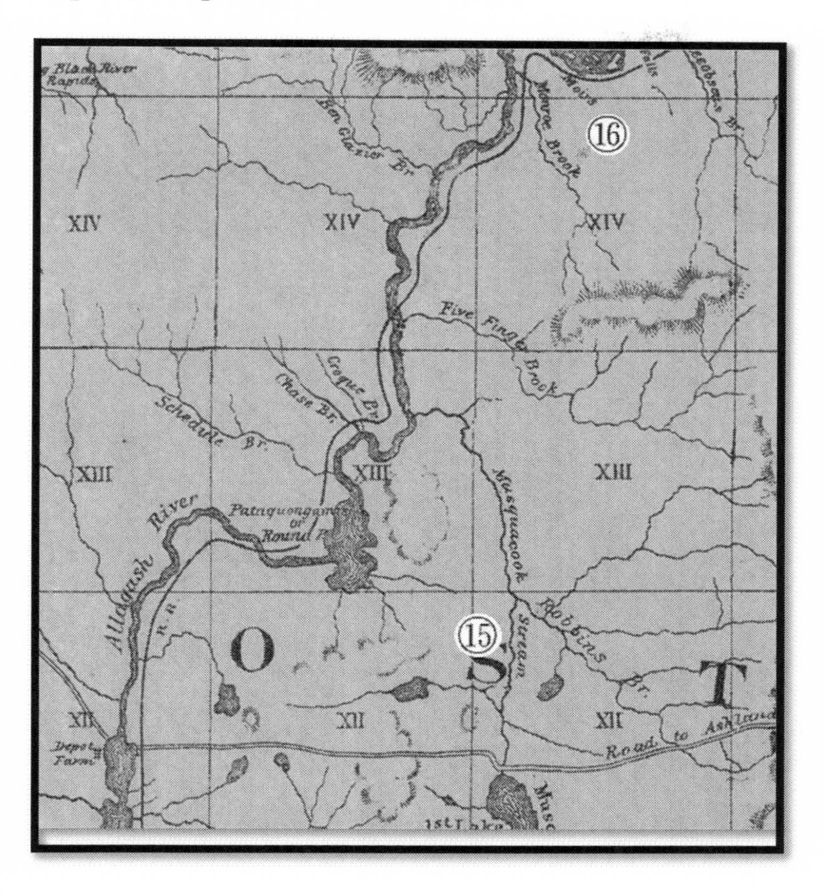

# Map 8 – Moirs to Edmundston

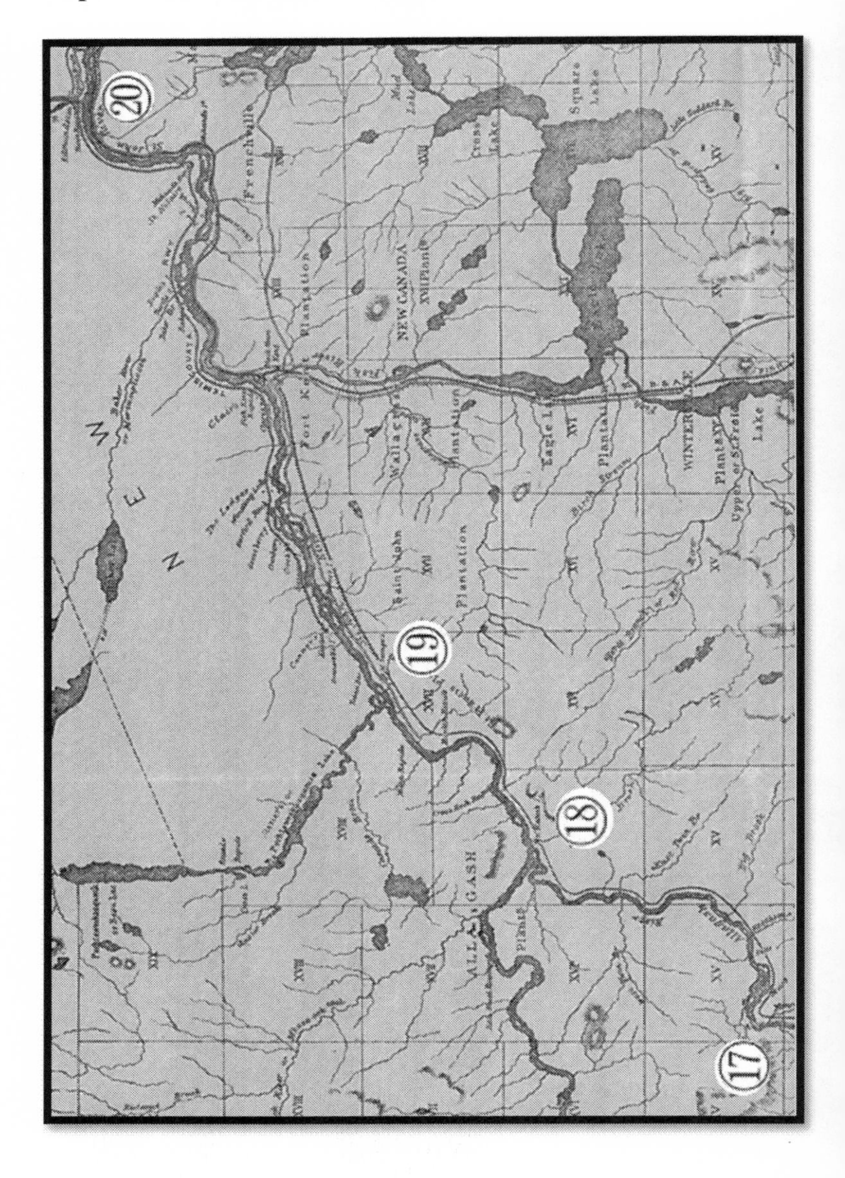

# Acknowledgements

If it were not for Frederick Wilcox having such high praise for the original Hubbard text, I may never have discovered this treasure of a book. I am grateful for his insight, conversations, and contributions to the re-issue of this American North Woods classic. I also appreciate Fred's contribution of his photo of Kineo rhyolite, and adding facts for the footnotes. Most of all, I am moved by his gift of entrusting to me his copy of the original book.

There was a second Wilcox that assisted me with this project. Thank you, Carol Wilcox, for your review of the introduction materials and fact checking on Hubbard's biography.

I thank my wife Meredith, who looks over my typing and supports my writing hobby in all that she does for me.

I acknowledge the Maine Historical Society in Portland, Maine and the Brown Research Library for access to their archives.

Lastly, I thank the Moosehead Historical Society & Museums for answering questions during the editing of this volume.

# About Frederick Thomas Wilcox

*

Fred grew up in the vicinity of Warwick, New York. He describes his childhood as one spent mostly exploring the woods and surrounding areas. Following graduation from Warwick High School in 1957, he attended the University of Maine where he studied Forest Management. Following graduation from the University in 1962, he moved to Pennsylvania to work as a forester for the Commonwealth of Pennsylvania. After a long successful career of 39 years he retired in 2001.

Fred currently resides in Hershey, Pennsylvania and still spends time exploring the woods and surrounding areas. He still includes trips to Maine in his vacation plans.

*Fred next to Eastern Hemlock in Cook Forest State Park, Pennsylvania*

*

## About Tommy Carbone

*

Tommy Carbone lives in Maine with his wife and two daughters. He studied electrical engineering and earned a Ph.D. in engineering management.

He writes fiction from a one room cabin, on the shores of a lake, that is frozen for almost six months out of the year, and moose outnumber people three to one. (I bet you can guess where that is...)

His first novel, "***The Lobster Lake Bandits – Mystery at Moosehead***," has made those 'from away' want to visit Maine. It's a big state – come explore.

# Books from Maine's North Woods

A Maine Novel

The second novel in the

***Moosehead Mystery***

series

**Hubbard's Guide**

to exploring

Northern Maine

**2020 Edition**

Made in the USA
Middletown, DE
07 April 2023